Nanomedicine

This book is an introduction to nanomedicine informed by a philosophical reflection about the domain and recent developments. It is an overview of the field, sketching out the main areas of current investment and research. The authors present some case studies illustrating the different areas of research (nanopharmacy, theranostics and patient monitoring) as well as reflecting on the risks that accompany it, such as unanticipated impacts on human health and environmental toxicity. This introduction to a fast-growing field in modern medical research is of great interest to researchers working in many disciplines as well as to the general public. In addition to an overview of the work currently ongoing, the authors critically assess these projects from an ethical and philosophical perspective.

Key Features

- Provides an overview of nanomedicine
- Employs a reflective and coherent critical evaluation of the benefits and risks of nanomedicine
- Written in an accessible manner intended for a wide audience

Nanomedicine

Panacea or Pandora's Box?

Jonathan Simon and Bertrand H Rihn

CRC Press is an imprint of the
Taylor & Francis Group, an **informa** business

First edition published 2023
by CRC Press
6000 Broken Sound Parkway NW, Suite 300, Boca Raton, FL 33487–2742

and by CRC Press
4 Park Square, Milton Park, Abingdon, Oxon, OX14 4RN

CRC Press is an imprint of Taylor & Francis Group, LLC

Library of Congress Cataloging-in-Publication Data
Names: Simon, Jonathan, 1964– author. | Rihn, Bertrand Henri, 1955– author.
Title: Nanomedicine : panacea or Pandora's box / by Jonathan Simon and Bertrand Rihn.
Description: First edition. | Boca Raton : CRC Press, 2023. | Includes bibliographical references and index.
Identifiers: LCCN 2022043616 | ISBN 9781032435435 (hardback) | ISBN 9780367436247 (paperback) | ISBN 9781003367833 (ebook)
Subjects: MESH: Nanomedicine | Nanostructures—toxicity | Nanotechnology—methods
Classification: LCC R857.N34 | NLM QT 36.5 | DDC 610.28—dc23/eng/20230201
LC record available at https://lccn.loc.gov/2022043616 61

ISBN: 978-1-032-43543-5 (hbk)
ISBN: 978-0-367-43624-7 (pbk)
ISBN: 978-1-003-36783-3 (ebk)

DOI: 10.1201/9781003367833

Typeset in Times
by Apex CoVantage, LLC

Contents

Authors

Jonathan Simon, PhD, HDR, teaches philosophy of science at the Université de Lorraine (in Nancy) and is a member of the Archives Poincaré (UMR 7117). Trained in the history and philosophy of science at the LSE and in the Department of History and Philosophy of Science of the University of Pittsburgh, he has taught in the United States, Australia and France. His research work has been on the history and philosophy of pharmacy and chemistry.

Bertrand Henri Rihn, MD, DSc, professor of biochemistry and molecular biology at Université de Lorraine, France, is a leading expert in the field of particulate toxicology, with research topics including safety toxicology, mechanistic toxicology and immunotoxicology by investigating transcriptomic changes in macrophages following man-made particles and nanoparticle exposure. He is the head of the 'Nanomaterials and Health' team at the Institut Jean Lamour (Nancy, France), where he is initiating a new research field devoted to the study of the interaction (toxicity and biocompatibility) of macrophages and new implant alloys. In addition to his research, Professor Rihn has also worked as clinical laboratory investigator in bacteriology and virology and is Medical Officer-in-Chief at French Defence Central Health Service (Service de Santé des Armées) involved in medical and sanitary support of the French army. He was awarded the Baratz Award (2004) and the Taub Award (2011) from the French National Academy of Medicine and served as the president of the French Society of Toxicology from 2007 to 2009.

1 Introduction

Thirty years ago, nanomedicine*[1] barely existed (Weber, 1999). Today, it is making a tentative entry into an increasing number of areas of healthcare and health maintenance. While not yet an everyday reality for most of us, it promises to become one over the next 20 years (Mittal et al., 2022). This book is about the broad domain of nanomedicine as it exists today, but it aims at being more than just an introduction to this already diverse and constantly expanding field. While we will be presenting a sample of the wide range of projects that fall into this category and asking what they can potentially offer in terms of human health, our principal goal is to use these illustrative examples to explore the wider issue of how nanomedicine promises to change the diagnosis, treatment and general management of disease. It is important to remember that nanomedicine is not just about innovative treatments for disease, but it also opens up new prospects for how we diagnose disease, manage our health and how we will live our lives in the future.

One lesson that the history of science in general and the much shorter history of the nanosciences in particular teaches us is how difficult, if not impossible, it is to predict how this scientific field will develop over the medium and long terms. Further, as we will argue toward the end of this book, it is not even clear whether nanomedicine will still constitute a term of reference in a decade's time. But while the field may change, the techniques and materials introduced through nanotechnology and the nanosciences will become not only more numerous but also more present in our lives. What then will be the value of these novelties introduced through the exploration of the nano? This too is uncertain, and we have chosen the subtitle 'panacea or Pandora's box' to evoke the extreme limits of the life-changing potential of nanomedicine. At one end of the spectrum, the *nano* approach to health may allow us to treat if not cure a number of recalcitrant diseases, particularly chronic diseases associated with old age, like cancer, Alzheimer's disease or Parkinson's disease. While antibiotics and the many other innovations in medicine and hygiene introduced in the course of the twentieth century have drastically diminished the impact of deadly infectious diseases and even eliminated some of them all together, chronic diseases, notably cancer, have proven more difficult to overcome. Is nanomedicine the next step forward in terms of eliminating these persistent health problems?

As far as cancer is concerned, nanomedicine has the potential to give physicians a strategic edge in the combat against the deregulation of the patient's own cells, and may be able to transform situations where palliative care seems the only option into ones where doctors can propose an effective treatment. Curing cancer is a laudable but notoriously elusive goal. We will discuss the prospects of this kind of research in the first section of this book, where we treat nanopharmacy. Cancer will also play a prominent role in our analysis of theranostics, techniques combining therapy

[1] The terms marked with an asterisk are featured in the glossary at the end of this volume. This being said, the meaning of the term nanomedicine is one of the central objects of the present book.

DOI: 10.1201/9781003367833-1

with diagnosis (Wang et al., 2022). Effective treatments for cancer, early diagnosis of chronic disease, real-time monitoring of a patient's health, and the tailoring of treatment to the patient's own genetic makeup—all this is nanomedicine as panacea. At the other end of the spectrum, the reference to Pandora's box evokes the potential dangers of working at the nano level in terms of both human health and the wider environmental impact of nanoscience. Pandora, the first woman according to Greek mythology, was instrumentalized by Zeus to punish mankind; she opened the box containing all the evils, including disease and death, releasing them into the world. When she finally managed to close the box again, only hope was left inside. Is this what we are doing by embarking on the industrial manufacture of nanomaterials or their widespread use in pharmaceuticals and cosmetics? Do we risk introducing insufficiently understood and tested substances into our bodies and into the environment? Might these sub-microscopic materials cause much greater harm than the benefits that they seem to offer? An exploration of nanotoxicity will map out some of these new threats to human health brought along by the nano. While the reality of the impact of nanomedicine (and the nanosciences more generally) is liable to fall in between the two extremes represented by Pandora on the negative side and the panacea on the positive, bearing these limiting horizons in mind will help us to put the various developments we will be considering here into perspective.

1.1 WHAT IS NANO?

The prefix nano in the word nanomedicine signals the domain of the very small. Invisible is only the beginning of the description when you get down to this scale. Take a single strand of hair; it is around 50 μm thick or 0.05 of a millimeter, and is already quite difficult to see. A nanometer is one thousandth of a micrometer (a millionth of a millimeter or a billionth of a meter) and therefore, thinking back to the single strand of hair, it is easy to understand that objects need to be tens of thousands of nanometers wide just to be visible to the naked eye. Scientists have known for well over a century that at the scale of the nanometer we enter a very different world of atoms and molecules; a hydrogen atom, for example, is about one-tenth of a nanometer in diameter, and the much larger, but still invisible glucose (sugar) molecule is about 1.5 nm long. Over the course of the twentieth century, physicists discovered that the laws governing the behavior of matter at this scale, quantum mechanics, are quite different from those that are used to describe and predict the macro world in which we, as human beings, are accustomed to operate. While the juncture between these two worlds remains a scientific and philosophical conundrum, results achieved at the nano level confirm the 'exotic' properties predicted by quantum mechanics, including the superposition of states, and an inherent uncertainty in the determination of physical parameters like position and momentum simultaneously. Thus, working at the nano level automatically brings into play these quantum properties, opening up possibilities that appear not to be available at larger scales.[2] This being

[2] There is, after all, no reason not to believe that all matter and its interactions answer to the laws of quantum mechanics and so the superposition of states and the dependency of velocity and position on measurement apply, albeit to a suitably negligible degree, to everything.

said, as we shall see in the following chapters, perhaps the greatest effect in terms of the medical use of nanoparticles is the huge increase in surface area that nanoparticles present when compared to bulkier materials.

The origins of the nanosciences, the field that has provided the conceptual and material tools behind today's nanomedicine can be traced back to the closing decades of the twentieth century. The development of the scanning tunneling electron microscope (STEM)* stands out as a landmark in this history, marking the beginning of the modern nanoscience adventure. The invention of the STEM in 1981 by Gerd Binnig and Heinrich Rohrer at IBM gave scientists a new form of access to the nanoworld of atoms and molecules. Although a microscope by name, the STEM is a long way from the typical light microscope that most people have used at school to look at plant and animal cells; it not only allows scientists to 'see' objects and surfaces molecule by molecule (in the sense of providing images) but also enables them to move single atoms from one place to another on an appropriate surface. The nanosciences, therefore, have opened up the possibility of intervening with precision at the molecular and even atomic level.

Over the last four decades, the STEM has found a range of applications, allowing the activation of rudimentary nanomachines as well as the precise characterization of a wide variety of molecular structures. Despite the importance of the STEM in the development of the nanosciences, much of the research in this area uses techniques that were around long before the invention of this instrument and were considered part of chemistry or microphysics before being transferred to the nanosciences. This transfer of techniques from other well-established scientific disciplines, like chemistry and physics to the burgeoning nanosciences, and in particular their applications in nanomedicine, will be another recurring theme in this book. The relationship between the nanosciences and the better established scientific disciplines like chemistry and physics remains an intriguing question for the philosopher and the professional scientist alike. In the 1980s and 1990s, there was a feeling that within a few decades the scientific world would be populated by nanoscientists, and that chemists, physicists and maybe even biologists would simply fade away as categories of scientist; but this shift never took place, and it looks unlikely that it will in the near future. Indeed, today, it is rather the future of the nanosciences that is uncertain. Nevertheless, for the moment the nanosciences remain the domain of physicists and chemists whose primary allegiance is to these historic scientific disciplines. In the course of this book, we will ask once again whether, 30 years into the history of the nanosciences, there is a major disciplinary shift on the horizon that might carry a new generation of scientists into nanomedicine. And if this is the case, will nanomedicine occupy the same disciplinary position or niche as it does today?

While we start our enquiry with nanopharmacy, as perhaps the most evident application of nanoscience to the health sciences, we will expand from here to consider a number of other areas that fall under the heading of nanomedicine for databases of publications, such as the American virtual library of research publications in medicine provided by PubMed. Even with this strategy, however, we could not claim to have covered all the various techniques and materials that fall under the head of nanomedicine. Indeed, we are only too aware that a book on nanomedicine written just five or 10 years from now could very well look quite different.

1.2 THE STRUCTURE

The book is structured around seven different areas of nanomedicine that we will treat in separate chapters. Nanopharmacy, diagnostic techniques (including theranostics—the combination of a diagnostic technique with the treatment of a disease), health surveillance (monitoring through microarrays and probes), genetics, nano-toxicity, organs on a chip and reconstructive surgery. In each of these chapters, we will explore a few carefully chosen examples to convey the spirit of the work currently being carried out in the domain, and in each case we will be asking what are the real shifts this work represents in terms of the scientific approach to medicine and human health. The aim is to identify credible trends in order to see where this research might be taking us. Thus, we will be aiming to throw light on what are the real innovations against a much more general background of excitement around a multitude of research projects, many of which will in all likelihood end up leading nowhere. This is probably the hardest task we will set ourselves, because, while it is easy to say that we want to consider real innovations that will change modern healthcare and not be seduced by the hype that has developed around certain projects, this is very difficult to do, and we are as likely as anyone else to make errors in our judgment of what is a significant development in nanomedicine, or a promising line of research, and what is not.

1.3 THE PLACE OF NOVELTY IN INNOVATION

As with many other scientific and technological innovations, it would be easy to argue that there is nothing fundamentally new about the nanosciences, and so presumably nothing new about nanomedicine itself. The argument here would be that the innovations brought about in the field do not represent anything authentically novel, but rather a repackaging of (or more positively a reconfiguration in) approaches and concepts that have already been applied in other domains. This is a common trope in the history and philosophy of science and technology; what is presented as a radical innovation, such as the steam engine, the telephone or the Internet turns out, upon closer examination, to be better understood as a new version of a phenomenon that has been seen before; the Internet, for example, can be understood as being fundamentally another communications revolution like the introduction of the telegram or printing technologies, but in a new form; while it is revolutionary in a sense, in another it can be considered to be just an extension of other related, but more fundamental revolutions that took place earlier. That is to say, the apparent novelty of the nanosciences can be considered a quantitative and not a qualitative transformation with respect to the science and technology that preceded it, and so there is nothing essentially new here. At the same time, how many times have we read over the last 25 years that the nanosciences represent a radically new way of approaching the world that is likely to change many aspects of our lives and those of our children? As a cross-disciplinary scientific domain that brings together a host of different specialties, the nanosciences propose a conceptual perspective that opens up new and exciting technological horizons. Is this enough, however, to qualify the field as an authentic scientific novelty? This is one question that we will at least be addressing in this book. So which is it? Radical novelty or a new version of something that we have seen before? While we will be discussing this issue, we do not have the pretention of providing a definitive answer.

1.4 THE DEFINITION OF NANOMEDICINE

There is an obvious question with which we can start, and that is: what is nanomedicine? In the most general terms, nanomedicine is the application of the nanosciences and nanotechnology to human health, which, since the emergence of the nanosciences at the end of the twentieth century, has been a constantly growing and diversifying domain. Nevertheless, despite its apparent clarity, this definition does not really help us, as it does not tell us what should or should not be included under the heading of nanosciences and nanotechnology (we will refer to both of these areas together as simply 'nano'). If the nano is conceived of as the production of new materials at the molecular level, then the nanotechnologies have clearly already changed the material content of our world. Nanotubes and graphene are just two examples of innovative materials that have been characterized and produced in industrial quantities as a result of research in the nanosciences and its application as nanotechnology. This being said, polyvinyl chloride (PVC) and nylon are another two examples of novel artificial materials that do not occur naturally but are now abundantly present on our planet. Developed as industrial products by chemists in the period leading up to World War II, these polymers have now been mass-produced for almost a century, meaning that their conception and production pre-dated the nano-revolution by at least 50 years. Why should this polymer science not be considered nanoscience? One pragmatic argument is that if we include the science of polymers as nanoscience, it becomes difficult to maintain the idea that the nanosciences represent a new domain. PVC, after all, was first produced toward the end of the nineteenth century and mass manufactured in the decades that followed, so well before the introduction of the scientific domain known as the nanosciences.

1.5 CAN THE DEFINITION OF THE NANO (NANOSCIENCES AND NANOTECHNOLOGY) BE LIMITED TO SIZE?

At least for the moment, the nanosciences constitute a well-identified domain for research and publication. A widely used definition of the nanosciences states that they are the sciences that deal with objects or substances in the 10 to 100 nm range and possess a high surface area. While this definition has the advantage of being clear and categorical, it has the drawback of including a number of practices, as we have already suggested, that existed long before anyone ever talked about or even thought about the nanosciences. Take the example of chemotherapy against cancer, where chemical compounds are used to treat a cancerous condition. A chemotherapeutic agent like cisplatin, for example, is a molecule that falls within the 10 to 100 nm range, and yet it was the product of traditional synthetic chemistry, initially produced by chemists in the middle of the nineteenth century, long before the techniques that founded the nanosciences had been developed. We will return to this conundrum in the first section of the second chapter on nanopharmacy, where we will see what difference there is, if any, between the 'traditional' approach to cancer chemotherapy and the different applications of the nano in this domain. Toward the end of the book, we will explore what is currently known about the hepatitis C virus, in order to challenge the arbitrary nature of any such quantitative cut-off point between the nano and the non-nano.

The main aim of this book is to provide a clear and comprehensible presentation of what nanomedicine is and how it promises to change healthcare in the future. To do this, we combine the different academic orientations of the two co-authors. One is a medical doctor specialized in nanotoxicology, and the other a philosopher of science who has worked extensively on the history and philosophy of pharmacy. Together we hope to provide a selective overview of nanomedicine that is both accurate and accessible. While doing so, we aim to raise questions about the domain that deserve serious reflection on the part of both researchers working in or around the field, and a non-scientific public who wish to exercise their critical reflection on the domain. At times the writing might seem rather technical and inaccessible, and at others, too simplistic. Overall, we aim to provide accessible descriptions and, more importantly, philosophically informed conclusions, more often than not in the form of questions that need to be asked and possible problems (particularly with respect to the potential toxicity of certain nanoparticles or nanomaterials) that need to be addressed in a proactive way.

While it seems evident what an academic training and experience in toxicology can bring to the interrogation around the nano in medicine, it is perhaps less obvious what the philosophy of science can contribute to this exercise. We hope that the combination of these two skills, which we consider complementary, will allow us to better identify what is original and interesting in the nano, and how this new domain will impact medicine and healthcare more generally. We will here be constantly trying to identify what is new and interesting in the developments identified as nanomedicine in order to give an accurate picture of how this domain is liable to develop. Of course, using current trends to identify future orientations in any scientific field is a perilous venture, but we hope to pick out the major significant trends in research over the coming decades. Philosophers tend to speculate about the future, perhaps more than scientists, or at least differently from them. Thus, on the side of predicting the future of nanomedicine, we will try to limit our speculation to what is reasonable, without offering any guarantee that our predictions will prove correct.

Although clearly one aspect of the book, we are not just presenting an assortment of nano projects in the medical and biological sciences. We also want to understand what they promise to change in healthcare. The experience of one of the authors in nanotoxicity will help to keep the presentation accurate, while the philosophical eye of the other will help to keep the larger questions in view and avoid losing sight of the wider narrative in the intriguing and often challenging details of individual research projects in nanomedicine, however ingenious and promising they might be.

As we have already mentioned, nanomedicine is a large and diverse area of research and one that is expanding in a number of different directions. We cannot, therefore, hope to cover every aspect of nanomedicine and, in practical terms, we have had to choose one of two options. Either we could try to cover the whole domain, necessarily superficially and partially, or else we could examine certain areas more closely, focusing on a limited number of projects to illustrate the research in the chosen areas. Opting for this second option, we still provide an overview which will allow us better to understand the nature of the field of nanomedicine as a whole, but means that the treatment will be far from exhaustive.

The tentacular nature of the nanosciences and nanotechnologies with their numerous ramifications that promise to reach into every area of our lives makes it an interesting subject for study. The nano are evolving fast, and nanomedicine reflects this expansion in its own terms. While we might initially have limited our idea of nanomedicine to new medical drugs designed by nanoscientists for specific therapeutic purposes, there is probably more research today on micro-arrays and other nano-based diagnostic tools than on this sort of 'traditional' nanopharmacy. We will do our best to give an accurate portrayal of this field even while it is in the process of evolving and transforming and maybe, even, as we will discuss at the end, nanomedicine might end up disappearing as an independent category for scientific research.

1.6 TO GO FURTHER

For a general introduction to nanomedicine from almost a decade ago, see Astruc (2015) and Chang et al. (2015). For a contemporary view of the future development of the field, see Richardson and Caruso (2020).

Astruc, D. (2015). Introduction to nanomedicine. *Molecules* (Basel, Switzerland), *21*(1), E4. https://doi.org/10.3390/molecules2101000 4

Chang, E. H., Harford, J. B., Eaton, M. A. W., Boisseau, P. M., Dube, A., Hayeshi, R., Swai, H., & Lee, D. S. (2015). Nanomedicine: Past, present and future—A global perspective. *Biochemical and Biophysical Research Communications, 468*(3), 511–517. https://doi.org/10.1016/j.bbrc.2015.10.13 6

Richardson, J. J., & Caruso, F. (2020). Nanomedicine toward 2040. *Nano Letters, 20*(3), 1481–1482. https://doi.org/10.1021/acs.nanolett.0c00620

1.7 REFERENCES

Mittal, S., Chakole, C. M., Sharma, A., Pandey, J., & Chauhan, M. K. (2022). An overview of green synthesis and potential pharmaceutical applications of nanoparticles as targeted drug delivery system in biomedicines. *Drug Research.* https://doi.org/10.1055/a-18 01-6793

Wang, X., Sun, R., Wang, J., Li, J., Walker, E., Shirke, A., Ramamurthy, G., Shan, L., Luo, D., Carmon, L., & Basilion, J. P. (2022). A low molecular weight multifunctional theranostic molecule for the treatment of prostate cancer. *Theranostics, 12*(5), 2335–2350. https://doi.org/10.7150/thno.68715

Weber, D. O. (1999). Nanomedicine. *Health Forum Journal, 42*(4), 32; 36–37.

2 Nanopharmacy
What Is New With the Nano?

As we explained in the introduction, nanomedicine can be broadly defined as the application of the nanosciences to medicine, making it as limited or as extensive as nanoscience itself. The fact that the definition depends on the nature and extent of the nanosciences means that, thanks to the many technological innovations and new materials coming out of this field, we end up with nanomedicine being a surprisingly wide and even diverse domain. In this chapter, we will explore the use of the nanosciences directly in the production of new medical drugs, but in the chapters that follow, we will be considering other rapidly growing areas related to the monitoring of the human body, covering not only diagnostic techniques but also other approaches for tracking people's health, repairing damage to their bodies and keeping them alive and in good health for as long as is medically possible.

In the early days of the development of the nanosciences and nanotechnology, before researchers had even produced very many new materials such as the nanotubes and graphene that have since become familiar resources for scientific research and industrial manufacturing alike, scientists and futurologists were predicting all manner of developments, from the apparently plausible to the quite fantastic. In his book from 2006, which already set out to criticize the 'Nano-hype' of its title, David Berube described this period as one of 'nanomania', with scientists and social commentators making extravagant claims that were never likely to be realized (Berube, 2006). The nano was going to revolutionize every aspect of our lives. According to these projections, what kinds of innovation was nanoscience going to provide in the field of medicine? There was a lot of interest around targets for therapy; nanoscience was going to help identify and validate these targets, and this is clearly an ongoing project almost 20 years later. We will have more to say about the role of the nanosciences in targeting receptors associated with disease, particularly cancers, in our chapter on theranostics, although it should be noted that it is an orientation that lies behind many other projects in nanomedicine as well. Nanoscience was also going to revolutionize genetic screening, with laboratories on a chip promising the miniaturization of all the diagnostic techniques emerging out of genetics, and so our short- and long-term health status would be constantly monitored in real time. If anything did go wrong with our health (or even if it just became suboptimal), we would be informed about it, as would the appropriate health systems. In an ideal version of an integrated feedback system, the same nano-based technology that identified a health problem would also be able to resolve it. Predictably enough, the presentation of these possibilities with their vision of an invisible yet total surveillance of our lives fed into an anti-nano backlash, a movement that saw the nanosciences as another cog in the machinery of technological surveillance that governments or even private companies

DOI: 10.1201/9781003367833-2

were putting into place.[1] In the most extreme dystopic projections, these same companies would be the source of terrifying, invisible self-replicating nanorobots that could take us over from the inside, echoing similar fears about computer viruses or even human viruses that could lay waste to modern civilization. This wariness about the nanosciences has remained, although today the focus of distrust is more on issues around privacy and control over the data produced by nanosensors and other 'intelligent' nanoparticles and less about nanomachines running out of control, which was the stuff of horror stories at the very beginning of the century (Crichton, 2002). We will discuss these issues of privacy and the management of sensitive personal data in the upcoming chapters when we take up the issue of the possibilities offered by the nanosciences for monitoring individual health.

The global promise of the nanosciences has been summarized by adapting the old adage: 'Early to bed, early to rise makes a man healthy and wealthy and wise'. Likewise, the nanosciences will make every individual human healthier and wiser (and maybe wealthier as well, although this is less clear). In the future you will be able to live a longer, healthier life, not by ensuring you have a good night's sleep, but thanks to new scientific developments, with nanomedicine at the forefront. Here, the popular image is once again of nanorobots injected into the bloodstream that get to work repairing or even improving our bodies from within. Of course, the paradox is that this kind of projection can lead people to believe that they no longer need a good night's sleep at all and as a result they might end up with shorter, unhealthier lives despite having nanomedicine working for them. All medical progress, be it surgical, pharmaceutical or any other measures (whether therapeutic, diagnostic or imaging), shares this potentially unhealthy flipside, and it just seems to be constitutive of interventions on the human body that no benefits can be gained (in terms of immediate health or an increase in life expectancy) that are not accompanied by risks. While nano-based drugs will doubtless extend and maybe even save many lives, their undesirable side effects will probably also become visible over time. This is a theme that lies behind our consideration of the toxicological properties of nanoparticles in Chapter 6. But before we can reflect on this issue of the unintended, undesirable consequences of nanoscience in medicine, we first need to know how this recent scientific domain of nanomedicine is supposed to help us lead longer, healthier lives and, incidentally, make us more intelligent.

2.1 THE NANOMEDICAL STRATEGY

In the short term, the nanosciences will contribute to both sides of a two-prong global strategy aimed at prolonging healthy human life. The two complementary areas are those of treatment and diagnosis, which for most of the history of medicine—although undoubtedly related—have been conceived of separately, but more of this in subsequent chapters where we will look into the diagnostic dimension

---

[1] Probably the most vehement opponent to all things nano, at least in France, was the Grenoble-based movement '*pièces et main d'œuvre*' which in the 2000s disrupted consensus meetings and published arguments against all aspects of the nanosciences. www.piecesetmaindoeuvre.com/ (accessed 24 May 2022)

of nanomedicine. Incidentally, we can already note that there is an obvious third element that is missing from this global health strategy: prevention. As we shall see in the chapters that follow, the nanosciences promise to contribute to disease prevention as well, although on an individual rather than a collective basis, heralding a profound transformation in approaches to public health.

Let us consider, then, this crucial pair of treatment and diagnosis and how the nanosciences might change them both. The first prong of this strategy involves using nanosciences for the more effective treatment of disease, particularly the groups of diseases that have been on the rise and will doubtless continue to be the biggest killers in the twenty-first century: heart disease, neurodegenerative disease and cancers. The second prong will be the subject of the following chapters and involves prolonging life expectancy through widely disseminated diagnostic devices that will allow ordinary people (although, like space tourism or even vaccination against the COVID-19 virus, these techniques will undoubtedly benefit the wealthiest populations first) both to commit to healthier lifestyles and to be alerted to the presence of disease earlier. This approach will help across the board, notably for the greatest health challenge currently facing industrial societies, ensuring healthy aging in a rapidly growing population of older citizens.

Apart from speculating about what the future might bring, which seems impossible to avoid in any discussion of the nano, we want to concentrate on the developments that are already taking place, notably in terms of the vectorization of drugs for improving their delivery to appropriate targets in the human body. After a presentation of the overall lines of research in nanopharmacy, we will address the question of what nanoscience and nanotechnology have brought to the table that is genuinely new in this domain of medical drugs. We will argue that the innovative nature of the nano is less obvious in this area of the extension of drug research than it is in other fields like health monitoring, where nanomaterials and nano approaches more generally are changing the ground rules. These integrated health systems that mobilize nanotechnology will be the subject of subsequent chapters.

2.2 ISN'T EVERYTHING NANO?

We can start with an observation which is far from being original but is obvious to anyone who takes the time to think about the nature of the nanosciences; there have always been nanoparticles in the world, first naturally occurring ones, and later artificial nanoparticles introduced by humans. And this observation applies just as much to medicine, and in particular pharmacy, as it does anywhere else. Thus, a chemist can tell you that our everyday medical drugs, like aspirin or paracetamol, for example, are themselves nano-objects. The paracetamol molecule is less than 1 nm wide, and so, while a typical paracetamol pill is maybe 5 mm wide (laying them out side by side, you could, in principle, line up five million acetaminophen (paracetamol) molecules across the width of one of these pills), simply grinding up a Tylenol® pill into a fine powder can already bring us into the domain of the nanometer. But what is particularly interesting about the switch from macroscopic objects like a paracetamol tablet down to the scale of nanoparticles is that as particles get smaller, the properties of the substance itself can change. Thus, in its nano-form, a molecule will not necessarily have the same properties as it does as a pill that can be held between finger and thumb. Some inorganic molecules, like gold and silver, have been found to have medically interesting properties as nanoparticles that they do not possess in their usual form of large grain powders, let alone as silver engagement rings or as gold ingots.

One of the things that changes as particles get smaller is that the surface area in contact with the physical, chemical or, what interests us particularly here, biological environment grows exponentially. Thus, if you take a regular cube and divide it in half vertically and horizontally to give eight similar cubes with edges half the length of the original one, you have effectively doubled the surface area of the substance. If you keep on doing this down to the nanoscale, you end up with materials with very large surface areas. These enlarged surface areas are normally expressed in terms of square meters per gram of material, and nanoparticles frequently have a surface area in the range of 10 to 50 m^2/g. This means that just one gram of a nanoparticle can have as much as 50 m^2 of its surface in contact with the environment. More exposed surface area increases the possibility of contact with other molecules; this not only makes nanoparticles more reactive than other forms of chemical substances but can also lead to novel chemical and biological properties. Titanium dioxide, for example, has been used for a long time in industry as a colorant, but in its nanoparticulate form, it has been found to absorb ultraviolet light. This form of light, with a shorter wavelength than visible light (although not as short as X-rays), is the principal source of damage to the skin from exposure to sunlight, not just leading to sunburn but also potentially provoking life-threatening skin cancers such as melanoma. Thus, nanoparticles of titanium dioxide are now widely used as an additive in sunscreens and other skincare products making it the single most common nanoscale health product. This, has, of course, triggered concerns both about their potential toxicity for people who use these products and about their effect on the environment, once washed off the skin (Yuan et al., 2022).

In research aimed at improving the effectiveness of antibiotics, silver nanoparticles have been found to alter the permeability of cell membranes. This is a property linked to the size of the particles, and so associated with the use of the substance as a nanoparticle, and it is hoped that adding these silver nanoparticles to existing antibiotics will improve their anti-bacterial activity and maybe even overcome drug resistance, a significant problem confronting antibiotic therapy today (Hassiba et al., 2017; Whitehouse, 2015).

This kind of interaction between cells and nanoparticles opens up a wide range of therapeutic possibilities, as nanoparticles can penetrate where larger agglomerates remain blocked, allowing them to breach formidable biological obstacles for the circulation of drugs, such as the blood–brain barrier that protects the brain from infectious material present in the blood; we will see the full significance of this property when we turn to the topic of vectorization. Of course, this kind of promiscuity of the nanoparticles represents a health risk as much as it introduces new therapeutic potential, because many of these barriers, like the blood-brain barrier or the mucus barrier in the lungs, protect sensitive organs from dangerous microscopic intruders, notably bacteria and viruses. We shall see later in the book how the property of being able to cross this kind of biological barrier, which can be seen as a potential advantage in nanopharmacy, poses a particular threat to human health. Indeed, these kinds of consideration feed into the field of nanotoxicology, which is today a very active area of research. Nanotoxicology is regularly turning up surprising results suggesting that nanoparticles behave differently from microparticles, sometimes in quite unexpected ways. But more of this in the chapter dedicated to the subject of nanotoxicology.

2.3 TITANIUM AND OTHER NANO-METALS

The three most widely produced nanoparticles in the world are titanium dioxide (TiO_2), silicon dioxide (SiO_2) and zinc oxide (ZnO) but they are generally not used for medical purposes. As we just mentioned, titanium dioxide—in its nanoparticulate form—is added to sunscreen to protect the skin from harmful ultraviolet radiation, but it has also been found to have anti-bacterial effects, meaning that some of the millions of tons produced every year can be used in the fight against the infection of wounds.[2] Gold and silver nanoparticles appear to be even more promising in terms of healthcare uses than the mass-manufactured oxides mentioned earlier. In combination with antibiotics, gold and silver compounds in nanoparticle form help once again to kill bacteria. As we shall see in the next chapter, gold nanoparticles also enhance the effects of radiotherapy when placed in and around cancerous tumors.

2.4 NANOPHARMACEUTICALS BY DESIGN

Innovative nano-forms of metals and their inorganic compounds like those we have just been discussing represent a quite novel approach to pharmacy, and it is not unreasonable to expect other such nanoparticles to contribute to medical advances in the future. This kind of 'crossover' from the development of nanomaterials into the medical domain is a feature of nanomedicine that offers a great deal of potential. As we will see in the chapter on regenerative medicine, a number of nanomaterials developed in quite different contexts are being usefully employed for this medical specialty. In cases like these, the 'crossover' use of these materials is rarely something that can be predicted in advance. Thus, we will not spend more time on these nanomaterials with medical applications, but instead consider the direct impact of the nanosciences within pharmaceutical research. By looking at several examples of innovation in this domain, in particular the growing number of nanovectors used for drug delivery, we will be able to see what they have to offer modern pharmacy and how they might be able to change the nature of drug treatment in the coming decades.

Before presenting these nanovectors, we need to offer a word of warning to the readers about the changes that are likely to be brought about by these nanomedicines over the coming decades.

Born into an ambient discourse of revolution and renewal of the sector, it is very difficult to gauge the real impact of these drugs. Nevertheless, although their clinical impact might be minor at the moment, and much of the noise around them is more

[2] The US manufactures over a million metric tons of Titanium Oxide per year, although China produces more than twice as much, American Chemistry Council, 2018. *Elements of the Business of Chemistry*.

to do with investor interest than with increased life expectancy for patients suffering from diseases, it would be a gross underestimation of the importance of these approaches to say that nothing has changed. Thus, at the end of this chapter we will draw several conclusions about what the nano has contributed to pharmacy, aiming to strike a balance between critical cynicism and overzealous optimism, another incarnation of the opposite poles of Pandora's box and the panacea of the title of this book.

2.5 THE BOTTOM-UP PARADIGM OF MOLECULAR DESIGN APPLIED TO DRUGS

One promising application of the nanosciences to medicine is the transfer of the techniques developed in the production and manipulation at the nanometric or molecular level to the domain of pharmacy, renewing the arsenal of medicines for the twenty-first century. The new technical potential introduced by the nanosciences gives researchers the possibility (at least in principle) to precision-build molecules that could and should be better drugs by design. Putting a molecule together atom by atom, scientists could be sure that it would be perfectly conceived and precisely constructed for carrying out a very specific therapeutic task. Beyond providing new techniques for manufacturing active ingredients that have already been established as effective, engineering at the atomic level could generate new drugs that are conceived using computer modeling before being purpose-built from the bottom up. These drugs could well be more effective than the ones currently used in the treatment of disease, which are produced by chemists using top-down techniques, such as the purification of active ingredients in plants or their direct chemical synthesis. Although we are accustomed to think of the pills we buy at the pharmacy, all uniformly white, red or blue, each sealed in its individual wrapping, as the height of chemical purity, producing the molecules by chemical synthesis is relatively imprecise when compared to nanoscale synthesis atom by atom using the various nanotechnological methods that are continually developing and improving. The generalization of nanotechnological production promises not only to increase the purity of the active principles in our pills or injections, it also offers the possibility of reducing the energy needed for their production, a small but appreciable contribution to the fight against global warming. Additionally, drugs made from the bottom up should have fewer undesirable side effects, or maybe none at all. This idea of bottom-up synthesis, building up molecules atom by atom, was one of the leitmotifs of the nanoscience dream or fantasy in the 1980s and 1990s, but for the moment it remains a long-term perspective, except in a few cases, such as the synthesis of short strands of RNA and DNA as we shall see in later chapters.

In the end, however, this perspective of the bottom-up synthesis of 'designer' molecules as therapeutic agents has been eclipsed, at least for the time being, by that of vectorization in general and more specifically, the approach involving encapsulation of established drugs in man-made nanomaterials. As this is currently the most advanced field in terms of the practical application of the nanosciences in pharmacy, we will present this area of research in more detail and assess what it is bringing to the table in both theoretical and practical terms.

In order to understand how nanopharmacy might provide new more effective drugs, first through molecular engineering and then by means of vectorization, it will help to understand the logic behind twentieth-century pharmaceutical research and development. This is best done by taking some time to think about the central idea of the 'magic bullet' and how, along with the evolving study of receptors, this concept has guided pharmaceutical scientists throughout the twentieth century and into the era of the nanosciences. These same nanosciences are in the process of renewing this magic bullet approach, taking an idea dating from the end of the nineteenth century and bringing it up-to-date with the aid of the unprecedented precision provided by the tools of this new domain.

2.6 FROM EMPIRICAL PLANT-BASED RECIPES TO INDUSTRIAL PRODUCTS: USHERING IN A NEW ERA OF CHEMOTHERAPY?

For most of the history of humanity, the drugs that people have used to treat disease have been extracted from the leaves, roots or flowers of plants, or the bark of trees. One only needs to browse through a pharmacopoeia from the seventeenth or eighteenth century to see that most of the medical drugs filling its pages are derived more or less directly from the plant kingdom. This plant-based pharmaceutical tradition was transformed in the twentieth century, thanks to a double development that combined to give what we would term the chemotherapeutic revolution. First, thanks to the analytical techniques developed in organic chemistry in the early nineteenth century, scientists were able to identify and characterize individual physiologically active chemical compounds. Later, with the discovery of a number of key reactions behind modern synthetic organic chemistry, scientists could synthesize these compounds, either directly from chemical raw materials or starting from readily available intermediary compounds. Like much of Europe's working population a century earlier, pharmacy moved from the fields into factories.

Let us consider first, the purification of the active principles of medical drugs, which started early in the nineteenth century. A group of such active compounds, collectively known as 'the alkaloids', provide the earliest and perhaps the best example of this trend. The first of these alkaloids to be isolated in a crystalline form was morphine, the physiologically active chemical compound produced by poppies and responsible for the narcotic (and addictive) properties of opium. While we do not have the space here to go into the long and interesting history of opium use, suffice it to say that by the early nineteenth century it had become a widely used medical drug, with its darker history of addiction and illegal use still in the making, particularly in Western countries.

In 1804, the German pharmacist Friedrich Sertürner managed to isolate a pure form of morphine, the alkaloid found in opium and responsible for its narcotic properties. To confirm the potency of this physiologically active compound, Sertürner did not hesitate to test it on himself and his students. In 1818, the French pharmacists Pelletier and Caventou crystallized pure strychnine from a South American plant already widely used as both poison and tonic, and in 1820, perhaps the most important medical discovery of the time, this same pair of French pharmacists prepared pure samples of quinine, the active principle of the bark of the cinchona tree used to treat malaria (Simon, 1999).

The obvious advantage of working with a purified organic molecule like quinine, rather than the diverse extracts of cinchona bark that were available at the time, was the gain in precision with respect to the dose administered to the patient. Different varieties of the cinchona tree had different concentrations of quinine in their bark, and the concentration also varied with the seasons and the age of the tree, as well as the process of extraction that was used to produce any particular extract of cinchona bark. Once the molecule (quinine) has been purified, the origin of the drug (which variety of tree or which extraction process is used) is no longer pertinent for determining the appropriate dose, and a doctor can prescribe a precise volume or weight of the purified product without worrying about its provenance. Apart from this improved control over the dose, the purification of an active principle like quinine involved a reconceptualization of what constitutes a medical drug. Today, in line with the purification of the alkaloids, pharmaceutical scientists identify a medical drug with its active principle, usually termed a 'molecule'. To take an example that should be familiar to most of our readers, if you swallow a paracetamol pill (300 mg or even 500 mg) for a headache, you are implicitly accepting this vision of the drug as the active principle. First, you buy the paracetamol pills at the pharmacy in a pre-packaged form, and you would never think of asking the pharmacist where it came from. Although the pharmacist would be able to give you an answer if you asked for the name of the pharmaceutical company that had produced the pills (this information is provided on the packaging as well as in the mandatory information leaflet that accompanies all approved drugs), they would probably not be able to tell you in what country it had been manufactured, let alone in which particular factory. But it is not just to avoid an embarrassing episode with your pharmacist that you don't ask this kind of question. You do not ask because most people (and here we include the vast majority of medical professionals) believe that these details concerning its provenance make no difference to the product; what is important is the chemically pure active ingredient present in the pill. While it makes a difference if the pill contains 300 mg or 500 mg of paracetamol, it makes no difference whether this paracetamol came from the United States, from Europe, from India or from China. Two hundred years ago, a good or pure drug was one made from good or pure raw materials (usually the leaves or roots of plants, as we explained earlier) by a competent, if not skillful pharmacist, while today a pure drug is one where the active ingredient is present in a precise quantity in its chemically pure form. Furthermore, the pharmacist who sells a drug to a patient has usually had no part in its manufacture. Whereas before the Industrial Revolution, pharmacists produced and were accountable for the medical drugs they sold, today they neither produce them nor are held accountable for their quality or security, with their responsibility limited to the precise conditions of use (incorrect prescription or failure to identify a dangerous drug interaction).

The second element of the chemotherapeutic revolution also has its origins in organic chemistry, but this time in the domain not of analytic but of synthetic organic chemistry. Starting from the middle of the nineteenth century, pharmaceutical laboratories began to synthesize the active principles of the drugs that they sold. While the intervention of chemical manipulation is already evident in a purified product like morphine, the partial and then total synthesis of molecules has added another layer to this chemical revolution.

Synthesis of the active principles of pharmaceuticals—the molecules discussed earlier—has once again changed the conception of a medical drug, particularly in terms of our standards of purity, as pharmaceutical laboratories and the public alike now expect their medical drugs to issue from test tubes (or, more precisely from industrial chemical factories) rather than from the leaves and roots of plants. Thus, following on from the purification of the active principle, synthesis represents a further transformation of our conception of drug purity, which is now synonymous with quality.

This being said, the most economically and politically significant contribution of such synthesis was to liberate industrial production from the problems associated with the supply of exotic natural raw materials. Be this as it may, in terms of our conception of drugs, the implicit re-definition of expectations concerning high quality pharmaceutical products has been much more important. At the same time, synthetic organic chemistry enabled the production of new, similar compounds, based on those that had already proven themselves effective in practice.[3] Here we come closer to our concerns with respect to the potential contribution of the nanosciences to our arsenal of drugs. To illustrate this development, we can take a well-known example of innovation due to the application of synthetic chemistry from the nineteenth century.

In 1856 (when he was only 18 years old) William Henry Perkin produced the synthetic dye *mauveine*. A student at the Royal College of Chemistry in London, Perkin was experimenting with coal tar (the by-product of the manufacture of gas used to light the city), looking for a way to synthesize quinine. As Pasteur so famously remarked, chance favors the prepared mind, and rather than just throwing out the brightly colored product of his reactions, which clearly was not quinine, Perkins kept it and tested it as a dye. Thus *mauveine*, an aniline dye, was the first of a large number of artificial dyes generated by organic chemists that would transform the clothing industry, and the political economy of dyes. While Perkins discovered this first artificial dye, it was the German dye industry that capitalized on the discovery and, taking advantage of the growing number of organic chemists coming out of practically oriented chemistry schools around the different German states, multiplied these artificial dyes, thereby transforming the market for dyed cloth and, incidentally, the world of fashion.[4]

New dyes were generally produced by tinkering with the chemical structure of a known aniline* compound in the hope of generating a new color or changing other properties (in particular the 'fastness' of the dye or how well the color fixed to a given fabric like cotton or silk). Synthetic organic chemistry was applied to the domain of dyes to generate a new range of colors to use in the clothing industry and the same principle was carried over from the domain of dyes into that of drugs. Just as mauveine was the result of a search to replace quinine, heroin was the result of a similar kind of chemical tinkering aimed at replacing morphine. The approach, which has become a fundamental principle for pharmaceutical research, and is likely to

---

[3] This is the principle of the 'me-too' drug, so widely exploited by pharmaceutical laboratories in the twentieth century.

[4] Anthony S. Travis (1990). 'Perkin's Mauve: Ancestor of the Organic Chemical Industry'. *Technology and Culture*. 31(1): 51–82.

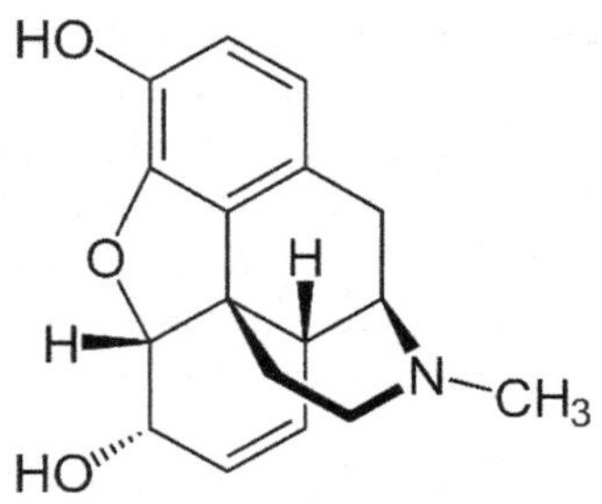

FIGURE 2.1 The structural formula of morphine. (From Wikimedia Commons, public domain: https://commons.wikimedia.org/wiki/File:Morphin_-_Morphine.svg.)

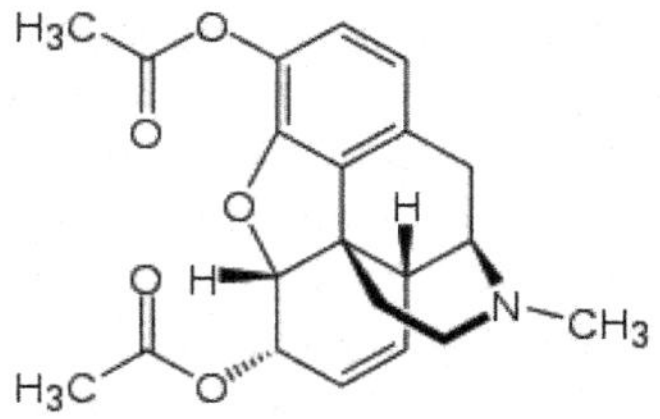

FIGURE 2.2 The structural formula of Heroin. (From Wikimedia Commons, public domain: https://commons.wikimedia.org/wiki/File:Heroin_-_Heroine.svg.)

be amplified in the nano era, involves exchanging chemical groups in the structure (replace a hydrogen atom with a methane group, for example) and then test the new variant to see the effects. The history of heroin is very informative from this perspective, as it can alert us to quite general opportunities and risks involved in introducing even minor structural alterations to physiologically active molecules. The synthesis of the heroin molecule (Figure 2.2) was intended, paradoxically, to provide a less addictive alternative to morphine (Figure 2.1), which was being widely used as a painkiller, but the organic chemists working on this project ended up producing an even more addictive product. Today, we would say that chemists and pharmacists explored the quantitative structure activity relationship (QSAR), testing the physiological functions of structurally related molecules in the hope of increasing the therapeutic activity as well as reducing undesirable side effects.

The second half of the nineteenth century was the period in which the practice of synthetic organic chemistry really took off. This rise in the synthesis of new compounds was fueled by industrial chemical concerns, with a new generation of organic chemists either collaborating with the large chemical dye companies like the Badische Anilin- & Soda-Fabrik (BASF) or Meister, Lucius & Co (later Hoechst) in Germany, or Wellcome in England or working directly in the industrial research laboratories that they set up.

2.7 PAUL EHRLICH, MAGIC BULLETS, RECEPTORS, AND THE LOCK AND KEY ANALOGY

Since the development of chemotherapy in the nineteenth century, which was, as we have just explained, tied into the large-scale production of new organic compounds by researchers working directly or indirectly for the burgeoning chemical industry, pharmacists and doctors have increasingly come to think of drugs as 'molecules', that is to say as a specific active ingredient. These molecules are meant to cure the disease without harming the healthy cells in the body: an ideal that has shaped our modern pharmaceutical world, and one that has fashioned our expectations, if not fantasies about nanomedicines.

In the early twentieth century, Paul Ehrlich, a gifted German chemist, physiologist and a founding figure in the fields of both immunology and chemotherapy, coined the term 'magic bullet' to describe his conception of an ideal medical drug. He was thinking in terms of diseases caused by bacteria like syphilis, and his magic bullet would be a chemical compound that killed the bacteria responsible for the disease while leaving the other cells in the body unharmed. By testing literally hundreds of new organic compounds against a model disease in an animal, he was able to identify an arsenic compound that could kill the bacterium responsible for syphilis. This drug was marketed by Hoechst starting in 1910 under the trade name of Salvarsan, as a specific treatment against syphilis. As might be expected in an arsenic compound, this drug had severe side effects, so while it was not really a magic bullet, its history exemplifies the direction taken by twentieth-century drug development. With this kind of magic bullet as their goal, and the associated scientific approach to pharmacy a leitmotif for advertising campaigns aimed at doctors, pharmaceutical companies screened the many thousands of new compounds produced by scientists in organic chemistry laboratories in order to try to find specific treatments with minimal side effects. As a complement to this investment in laboratory research, a new commercial model was put in place, promoting drugs using their scientific image, replacing the empirical approach behind the traditional pharmacopoeia with the modern image of the magic bullet conceived, designed and produced in the chemical laboratory.

This new model of pharmaceutical research was based on the principle of screening large numbers of molecules, albeit it in a strategic manner, using either bacteria grown *in vitro* or animals infected with the disease under study. This meant adding the chemical compound to a cultivated colony of the bacteria and observing the evolution of the colony for any anti-bacterial activity with the possibility of further testing the candidate drugs on laboratory animals infected with the disease. Initially performed by laboratory technicians, starting in the 1970s, these kinds of screening tests were first automated and then computerized. While this approach still represents a large part of the research effort deployed by the pharmaceutical industry, it has been increasingly challenged by 'rational' bottom-up approaches, notably based on detailed knowledge of the functioning of normal cells, diseased cells and the infectious agents still responsible for many of the most recalcitrant diseases.

'Magic bullet' is not the only term we have inherited from Ehrlich, as he also introduced the concept of receptors. In the same period, the chemist Emil Fischer introduced the analogy of the lock and key for thinking about the specificity of the action of enzymes. This last concept has been widely adopted and expanded in order to explain the specific action of medical drugs ushering in the era of quantitative structure activity relationship QSAR* studies. While Ehrlich's notion of receptors remained quite vague and was not clearly linked to any existing molecules found in the body, scientists' knowledge of receptors has since become highly specific and detailed, and today this knowledge of receptors lies at the heart of pharmaceutical theory and research, notably in terms of the mechanism of drug action. The lock and key analogy has been continually recycled in discussions about drugs and is used to explain the specificity of drug action as well as their possible side effects. The diffusion of the lock and key analogy is tied to the same period—the early part of the twentieth century—when researchers increasingly favored a specific view of disease, with bacteria and later viruses displacing more general dispositional theories. Over the course of the twentieth century, and particularly after the World War II, the lock has become identified with specific receptors, and so drugs that block or activate these receptors have become the embodiment of the key in the evolving paradigm of the magic bullet.

A good example of this kind of approach is the class of drugs known as the beta-blockers, a major pharmaceutical innovation dating back to the 1960s. In 1964, James Black, a gifted British chemist developed propranolol, the first beta-blocker that was marketed under the name of Inderal by the British company ICI (Imperial Chemical Industries). This drug was used to treat high blood pressure and works by binding to specific receptors in the heart—beta-adrenoreceptors, hence the name beta-blockers—meaning that neurotransmitters can no longer bind to these sites, blocking off a natural mechanism for raising the heart rate and so increasing blood pressure. Hence, the result of taking these drugs is a lowered heart rate and reduced blood pressure. With this kind of development, the lock and key metaphor had found its modern incarnation, with drugs as keys and receptors as the locks.

As we have already noted, receptors are at the heart of contemporary pharmaceutical models of drug action although the concept of the receptor has considerably evolved since 1900, when Ehrlich initially put forward the idea of a specific molecule involved in cellular immune reaction.[5] Today there are a wide variety of receptors, which might be any one of thousands of different proteins. Their chemical structure and biological function are both studied in as much detail as possible, as these receptors control the various exchanges between the cell and its environment, as much in terms of signaling as in the physical passage of molecules into and out of the cell. And in terms of pharmaceutical research, receptors are now the locks and medical drugs, the keys.

Today, receptors are very precise well-characterized chemical objects, which lead us to suggest that finding the key to fit one of these locks has become more than just a metaphor. The three-dimensional structures of receptors are modeled on computers, as are their interactions with a range of substrates or ligands, providing researchers with ideas for appropriate leads to develop new drugs. Zanamivir (Relenza®—initially developed in the 1980s by Biota Holdings, an Australian biotech company) is an example of this kind of drug, conceived with a very specific function in mind; it inhibits an enzyme called neuraminidase that participates in viral infection and

[5] Cay-Rüdiger Prüll (2010). 'Paul Ehrlich's Standardization of Serum: Wertbestimmung and Its Meaning for Twentieth-Century Biomedicine' in: Gradmann C., and Simon J. (eds) *Evaluating and Standardizing Therapeutic Agents, 1890–1950. Science, Technology and Medicine in Modern History*. Palgrave Macmillan, London.

reproduction, making it of potential interest in the fight against all sorts of viral diseases, including COVID-19.

2.8 NANOPHARMACY AS DRUG DESIGN

The lock and key model combined with current knowledge of receptors and the precision of nanoscale engineering means that the nanosciences hold out the hope of developing drugs tailored down to their last constituent atom to combat a disease through a specific intervention in the physiological function of a bacteria, a virus, a cancerous cell or any other kind of cell for that matter. Nevertheless, as we have already said, this kind of precision drug design remains a project for the future. For the time being, research and drug development mobilizing the nanoscale is focused on the domain of vectors rather than custom-built drugs.

2.9 VECTORS AND ENCAPSULATION

While this image of tailor-made drugs produced at the atomic scale using atomic force microscopes remains very seductive, looking at the lists of drugs classed as nanomaterials that have been approved for use in the United States, Europe and Japan gives us a more realistic picture of the direction that nanopharmacy has actually taken over the past decades. For the moment, this is not the highly specific design of new drugs outlined earlier, but rather the use of nanoparticles to deliver established drugs more reliably and more safely to their sites of therapeutic action.

Drug vectorization is a technique that places a physiologically active molecule (the active principle) within another larger molecule (the vector) in order to transport it at least into the patient's body, and preferably to bring it directly to the site of its therapeutic action. The idea is that this kind of targeted delivery will not only improve the efficacy—milligram for milligram—of the drug in the organism, but, by avoiding its dissemination throughout the body, it can also reduce the undesirable side effects caused by the presence of a highly physiologically active molecule elsewhere in the body.

Vectorization offers at least three advantages over the unmediated use of the drug that is being transported. First, the approach avoids the active principle being broken down or transformed by the body's immune defense system before being able to perform its therapeutic function, thereby increasing the effectiveness of an equivalent dose. This kind of dose reduction measure has implications for the safety of the drug as well, as lower doses normally translate into fewer side effects. Second, when the vector can be guided toward the affected organ or region, the local concentration of the drug reaching the diseased area is increased, which, following a dose–response logic should make the drug more effective. Last, but not least, this guidance of the drug delivery avoids its dissemination throughout the body in an uncontrolled way. Thus, the argument goes, with the drug being delivered exclusively to its site of action, the unwanted side effects will be reduced even further. The conceptual scheme is very appealing, and this vectorization approach is often presented as a major advance in modern medicine, particularly in the field of cancer chemotherapy. For the moment, it is hard to justify the claims that vectorization has brought about a revolution in drug therapy, but we cannot exclude the possibility that it will be seen to be one in the years to come. To give the reader some idea of how this area has developed over the last 30 years we will take them through the three generations of vectors that have already supplied drugs approved by the FDA (Food and Drugs Agency, the government body that regulates drugs in the United States) and other pharmaceutical regulatory agencies worldwide, meaning that many of these vectorized products are available for clinical use today. In our conclusion, however, we will be underlining the dominance of a relatively simple class of vectorized nanoparticles: ones that use albumin as a self-assembling vehicle for a wide range of drugs.

2.10 BASIC NANOMATERIALS: CHYLOMICRONS, MICELLES AND LIPOSOMES

The fact that the liquid inside our intestines is mostly water makes it difficult for us to absorb fatty nutriments. Water cannot dissolve fats directly, as you can easily observe yourself, if you ever have to wash the pans you use to fry up a greasy meal; the fat forms globules in the water rather than dissolving like sugar or salt would. As the interior of our intestines is a predominantly watery environment, this problem of the solubility of fats in water means we stand to lose out on the nutritional value of the fats in our food for just this reason. The body produces micelles known as chylomicrons to compensate for this problem by absorbing the fats, while they themselves are water soluble. Chylomicrons are naturally occurring micelles with a diameter of about 1,000 nm found in the hepatic portal circulation system (the blood flow between the intestines and the liver). They transport semi-digested fatty chemical compounds—triglycerides, phospholipids and Lipoproteins*—from the intestine to the liver for further digestion and are present in large quantities after a greasy meal.

2.11 MAN-MADE MICELLES AND LIPOSOMES

Imitating the structure and function of these chylomicrons, chemists have succeeded in synthesizing spherical micelles of between 20 and 250 nm in diameter without too much difficulty, because they are relatively stable structures. The particularly interesting feature of these spheres is that they are made of molecules possessing both hydrophilic (attracted to water) and hydrophobic (repelled by water) ends. With the hydrophilic ends facing outward and the hydrophobic ends facing in toward the center, the sphere can thus be charged with a similarly hydrophobic active principle and yet deliver it to its target as a water-solvent molecule.

Liposomes are larger spheres based on a similar principle but between 100 and 800 nm in diameter. They are formed by a bilayer of phospholipids and so can have an aqueous cavity inside them. The idea is that once introduced into the body, they can fuse with the plasma membrane to deliver either hydrophilic or hydrophobic drugs enclosed inside them directly to cells, thereby protecting the active principle itself from degradation. Thus, in the case of cancer treatment, a liposome vector with an active anti-cancer drug tucked inside can be carried by the circulatory system to a targeted tumor. The principle of using these molecules as vectors is quite straightforward; artificial liposomes, in the nanoscale range (around 100 nm), are loaded with a smaller medical drug. The combination is administered to the patient, and the active principle is more fully available for its therapeutic purpose than it would be administered on its own.

2.12 THE FIRST GENERATION—LIPOSOMAL VECTORS

The goal in preparing these liposomal vectors is to have a molecule into which the drug can be loaded efficiently, giving a system that controls drug delivery both over time and in terms of space, thereby increasing the availability of the transported drug in the body for a longer time. When used to treat cancer, for example, the idea is to increase the capacity of delivering the drug to the cancer cells and, overall, increasing drug efficacy while reducing toxicity, enhancing what is known as the therapeutic index.[6]

These initial liposomal vectors were mostly combined with anti-cancer drugs to see whether vectorization could provide more effective treatment with fewer side effects. This approach exploits certain features of cancerous tumors, notably the fact that they are well supplied with blood and relatively poorly served by the lymphatic system that drains used blood components away from the body's organs. The idea is that anti-cancer drugs protected from the body's metabolism will tend to accumulate in and around the tumors, making them more effective and causing less damage to healthy body cells.

As might be expected, most anti-cancer drugs are very aggressive, a reflection of the dangerous nature of the diseases they are used to treat, leading to high levels of undesirable side effects. This toxicity is in the very nature of cancer chemotherapy, as we can see by looking at an example of a commonly used anti-cancer drug that has benefitted from vectorization using liposomes; doxorubicin, a member of a group of anti-cancer drugs known as anthracyclines*. These anthracycline drugs act in a similar manner to kill cells, either by interfering with the mechanism of cell reproduction or by sabotaging cell repair mechanisms.[7] By disrupting the mechanisms of cell reproduction or cell repair at the molecular level, they can slow or even stop the spread of the cancer cells. The end result of their interference in cell function is that they trigger apoptosis* (the cell's own self-destruction mechanism), thereby reducing the size of tumors, the standard measure of their anti-cancer activity. This strategy of interfering with cell reproduction or maintenance is used to treat cancers (those involving solid tumors), as one feature shared by all cancerous cells is their capacity to reproduce very rapidly. Indeed, one definition of cancer is just this kind of cell reproduction running out of control with consequent malignant growth, making the disruption of cell reproduction a logical anti-cancer strategy. But cancerous cells are

---

[6] The therapeutic index is a measure of the relative safety of a drug weighed against its efficacy. It is the ratio of the amount of the drug that has a therapeutic effect to the amount that causes undesirable side effects (toxicity). Thus, a drug with a high therapeutic index has a large therapeutic effect with a low level of toxicity.

[7] The terms 'killing', 'interfering' and 'sabotaging' are clearly metaphorical, and, unfortunately, analyzing the use of these metaphors and what they can teach us about chemotherapy will take us too far from the focus of this chapter. Suffice it to say that the analysis of these metaphors can greatly help our understanding of cancer chemotherapy in particular and chemotherapy more generally.

not the only ones in the body to reproduce, and these anthracyclines also interfere with the reproduction of healthy cells, particularly the white blood cells that are produced in large numbers in the bone marrow to keep the immune system functioning. Thus, the use of anthracyclines is frequently accompanied by a condition called neutropenia, which is a reduced level of neutrophils, a component of our blood plasma that is essential to the immune system.

Myocet® is a pharmaceutical consisting of doxorubicin combined with a liposomal vector that received approval for use in Europe in the summer of 2000. The drug is used, but it is hard to evaluate its benefits compared to other drugs in use to treat cancer, particularly because of the prevalence of combination therapies to treat cancers. Although, it has to be admitted that with over 20 years on the market, this drug—and the similar vectorized anti-cancer agents that have followed—has not radically transformed the face of cancer chemotherapy, and the potential of these vectorized pharmaceuticals as the new nano-incarnation of the magic bullet has not yet been realized.

2.13 THE SECOND GENERATION—PEGYLATION

The developers of this first generation of liposomal nanovectors were concerned that these transport molecules or vectors were arriving at the threshold of the size that would be detected and destroyed by the human body's immune defense system. To escape attacks from the immune system, and the rapid elimination of the drug from the body, chemists proposed PEGylated* nanomaterials as alternatives to replace the first-generation liposomes. PEGylation* is the inelegant name for the addition of polyethylene glycol (PEG) molecules to the surface of the liposomes. PEGs are polymers* (chemicals made up of repeating units, in this case units of ethylene glycol) of varying lengths, and adding them to the liposomes makes it less likely for the molecules to be targeted by the immune system and broken down by macrophages or the mononuclear phagocyte system of immune defense, meaning that the vector and the drug it transports will remain longer in the body. The use of PEGylated liposomes to transport doxorubicin, for example, had the goal of reducing the drug's cardiotoxicity. PEGylation was thus a first step in a new 'stealth' strategy that started to be elaborated around these vectorized drugs. Indeed, stealth has become the leitmotif for the development of more sophisticated versions of these delivery systems, with this metaphor being added to the terminology of targeting that has always accompanied these nanodrugs.

Talk of magic bullets and targeting naturally makes the public think of war, and waging war on disease is a very common trope—particularly with respect to cancer. Thus, it is not surprising to see these kinds of metaphor transferred to nanomedicine (Oftedal, 2019). The more recent addition of the term stealth is consistent with this military lexicon, but the three metaphorical terms are far from having the same status, be it historically or conceptually. The magic bullet is a code word for participants and observers alike that draws their attention to the specificity of drug action and highlights the ideal of a drug conceived to fulfill a particular physiological function. As we have already explained, this has been a guiding model for drug development since at least the 1960s, with receptors as the focus for this approach, which brings us to the issue of targeting. Talk of targets extends the military metaphor but at the same time is more than just a transfer of the concept of a target for attack from the army to medicine, as it covers a much more general sense of what is being aimed at. Targeted therapy, like target audiences for cultural events, is as much about what parts of the body are left out of a treatment as it is about a goal to be reached. Returning to our example of vectorized doxorubicin, the decline in heart failure as an unwanted side effect in the switch from doxorubicin to Myocet or its equivalent, suggests how this targeting approach might prove particularly successful not because of where the drug goes but because of where it does not go. We will have more to say about targeting in the next chapter, as many of the theranostic techniques under development aim at improving drug targeting.

Finally, we get to stealth, which brings us back to an unambiguously military lexicon, evoking the stealth bombers built by the US Air Force at the end of the twentieth century. These airplanes were designed to escape detection by enemy defense systems, constructed in a way that rendered them invisible to the defensive devices in place, particularly radar. This is the logic of the most recent nanovectors that are being tested for the delivery of anti-cancer drugs, with the body's own immune system cast in the role of the enemy's defense.

2.14 THE THIRD GENERATION—NANOMATERIALS WITH SPECIFICALLY MODIFIED SURFACES

The evolution of these liposomal vectors has not ended with their PEGylation, as other chemical compounds have since been placed on the surface of the liposomes to perform a variety of different functions. As with PEGylation, one aim has been to reduce the elimination of the molecule by the patient's own immune system, thereby giving the drug contained in the vector more time in the body to act. Thus, vectors have been developed with cell membrane camouflage to render them less visible to the immune system. One example of this approach is the use of the CD47* receptor to trick the immune system, at least temporarily, and so allow the vectorized drug to remain in the patient's circulatory system for longer. The CD47 receptor bears the nickname of the 'don't-eat-me' molecule, as its presence on the surface of a cell serves as a signal to inhibit phagocytosis (Vandchali et al., 2021). Placing the CD47 receptor on the surface of the kind of nanoparticle vector we have been considering extends its lifetime in the body by using this signal to slow down, if not avoid the functioning of the body's immune system.

2.14.1 Multi-functional Vectors

This is the logic of 'stealth' described earlier, with various compounds placed on the surface of the liposomes to avoid triggering an immune response from the patient undergoing the treatment. But the additions do not end here, as chemists have also been able to place receptor-specific proteins and monoclonal antibodies on the surface of the liposomes, enabling the vectors to bind onto cancer cells, increasing the specificity of the drug action. Here, the historical reference that is often cited to explain the strategy is that of the Trojan Horse; the vector containing the anti-cancer agent is identified as 'friendly' by the cancer cells and is allowed to bind to it or even pass through the cell membrane where it can deposit its deadly cargo. These 'customized' vectors can also be equipped with magnetic functional groups, for example, which then allow external guidance to their site of action. Once again, we will have more to say about these methods of active targeting in the next chapter, as they are generally included under the head of theranostics. So we see how this third generation of liposome vectors can carry specific markers on the surface that allow them to bind to receptors on the targeted tumors. Of course, referring to receptors and talking once again about specificity brings us back to the ideal of the magic bullet that has guided drug research throughout its modern history. But how much closer to the ideal of the magic bullet do these third-generation liposomal vectors bring us? As these third-generation customized vectors are generally in the early stages of development and testing, it is hard to say what benefits they will bring to patients, particularly cancer patients in the coming years. But looking back across the history of these vectors in cancer chemotherapy that dates back to the turn of the century, it remains difficult to assess their impact at the clinical level.

In a note that accompanied its approval of Myocet (the liposomal doxorubicin product we presented earlier), the European Medicines Agency's Committee for Medicinal Products for Human Use commented that they were granting permission for its sale and use in Europe not because of any proven increase in effectiveness in treating cancer compared to standard non-liposomal chemotherapy but because of a lower risk of associated cardiac problems.[8] While this safety factor is a good reason to approve the drug for use, and to continue looking for alternatives, it does suggest that the conceptual innovation of vectorization has not translated into major advances in the effectiveness of chemotherapy, which was the initial goal behind their development. Of course, the advantage of running through several generations in this approach is that in case of a disappointing result at any particular stage we can always transfer our hopes onto the next generation. More generally, it is not hard to find presentations and pamphlets vaunting the theoretical virtues of these nanovectors, but it is much more difficult to find the results of clinical trials that prove their value in extending the lives of cancer patients. We are far from being the first commentators to signal the gap between the promises and the reality of high-profile scientific innovation, not just in nanoscience but in other areas of cutting-edge scientific research like genetics and nuclear fusion as well. Nevertheless, as in these other domains, while the faith put in the promises of the short-term benefits of the research was maybe misplaced, this should not blind us to its potential. Despite being unable to cure diseases as was promised at the beginning, genetics and, more particularly, genetic engineering have clearly changed our way of thinking about every aspect of life, including disease and health. Likewise, although nuclear fusion has yet to be put into operation in a useable form on this planet, if and when nuclear fusion comes on-line as a local energy source, it will completely change our relationship to energy production and consumption. Thus, even though the vectorized anti-cancer drugs on the market have not yet changed the prospects of those suffering from cancer, this does not mean that this approach might not hold the key to an effective treatment in

[8] www.ema.europa.eu (consulted 14 May 2020).

the future, if not a cure for certain forms of the disease, something 'classic' chemotherapy has already achieved.

2.14.2 Albumin as the New Nanoparticle Standard

We cannot bring this discussion of vectorization to an end without considering the rise to prominence of albumin-based formulations. Consulting the database of ongoing clinical trials, it is clear that albumin has become the protein of choice for manufacturing nanoparticle vectors for anti-cancer and even antibiotic drugs. So, what is the appeal of albumin?

Albumin is a protein that is found in the blood of mammals like humans reaching levels as high as 50 g/L. It is also found in high concentrations in the whites of eggs (around the yolk), which explains its name, as *albus* means white in Latin. Albumin is synthesized by hepatocytes and serves as a source of protein used by cells. Around five nanometers in diameter, globules of albumin possess hydrophobic cavities with the useful property of being able to bind non-ionized molecules, such as synthetic hormones, antibiotics or anti-cancer drugs. As early as 2000, albumin was shown to interact strongly with the anti-cancer drug paclitaxel (also known under its brand name of Taxol).[9] Astra Zeneca then developed nanoparticles of human serum albumin bound to paclitaxel as a new anti-cancer treatment (Abraxane®, Astra Zeneca laboratories), although generally used only in advanced cases.

The addition of saline solution to a lyophilized mixture of albumin and paclitaxel forms a nanoparticle of about 130 nm in diameter in a colloidal suspension. This process is very easy to carry out and the resulting albumin nanoparticle form displays greater anti-tumor efficacy than a similar dose of paclitaxel. This higher efficacy is probably due to the better uptake of the paclitaxel into cancerous cells once combined with the albumin, because the cancerous cells have higher needs for nutrients than their quiescent non-cancerous counterparts. This uptake is made by caveolin-1, a protein present on the membrane of numerous cancer cells, allowing accumulation of the drug vector in these cells. Once the albumin is hydrolyzed by the cell metabolism*, the drug is liberated and can block cell division by binding to a protein required for cell division. The interference with this cancerous cell division can stop the tumor from growing and spreading by metastasis*, also potentially exposing these cells to destruction by the immune system. Abraxane® is being tested in a large number of clinical trials as shown in Table 2.1.

As we can see quite clearly from this table, the dominant presence of albumin among clinical trials is due to Abraxane. Still the interest in this formulation will very likely place albumin nanoparticles at the top of these kinds of vectorized products, sidelining the more ambitious customized vectors discussed earlier (Larsen et al., 2016). In the end, the advantages of using a naturally occurring protein already found in high concentrations in the plasma might well outweigh the sophisticated 'stealth' tactics that formerly guided innovation in this area. Nevertheless, the fact that the dominance of albumin as a vector turns around a single drug means that this

---

[9] Taxol is an anti-cancer drug first extracted from *Taxus* genus tree bark or needles. Subsequently obtained by semi-synthesis, the fully synthesized form became available in 1994.

TABLE 2.1
Clinical trials involving nanoparticles (NP)

Nanoparticle-bound drug	Number of trials
Abraxane® (albumin-paclitaxel)§	4,639
Paclitaxel§	4,159
Abraxane® AND albumin§	952
Other NP of which:	478
Silver NP	28
Liposome (of which nanoliposome)	24
Magnetic AND iron oxide NP	23
SARS-CoV-2 NP	13
Carbon NP	11
Silica NP	6
Titanium dioxide	6
Zinc oxide NP	5
Hydroxyapatite NP	3
Polyethylenimine	3
LDL-like NP	3
Copper NP	2
PEGylated Liposomal Doxorubicin	2

As retrieved from https://clinicaltrials.gov in March, 2021; § we estimate that 16% of the studies retrieved with the keyword 'Abraxane' were conducted with paclitaxel alone, and not in its nanoparticle formulation, which would explain the discrepancies in the figures.

approach could easily disappear in case Abraxane is seen to be a poor investment, or is overtaken in terms of efficacy by a competitor in these advanced cancer conditions.

2.15 IS NANOPHARMACY A MEDICAL REVOLUTION?

If the introduction of the vectorized chemotherapy we presented earlier had led to widespread cures for cancers, or even a newsworthy leap in survival for advanced cancer patients, everyone would agree that the nanosciences had revolutionized medicine. But it is rare that major transformations take place like this, either in medicine or in science more generally. The introduction of penicillin followed by other antibiotics in the 1940s was one of the few cases in the history of pharmacy where a new drug radically changed the practice of medicine and the prospects of millions of patients over a short period of time. Similar arguments can be made for certain vaccines, which over time have served to eliminate diseases entirely from populations, the most notable example being the eponymous smallpox vaccine and its successors, which led to the eradication of the deadly disease, officially endorsed by the World Health Organization at the very end of the 1970s. Many other drugs with a significant impact on human health have been introduced before and after penicillin but without their global effect on health becoming clear until sometime later. When he proposed a model for scientific revolutions in his philosophical best seller from 1962, *The Structure of Scientific Revolutions*, Thomas Kuhn argued that it is only possible to identify a scientific revolution retrospectively after scientists have adopted a new paradigm and entered a new period of scientific stability. Only when they have returned to practicing what he termed 'normal' science based on a new paradigm to guide their research can they recognize the significance of the change that had taken place. Although we have already warned against the excessive speculation that has accompanied the nanosciences and nanotechnology since their inception, the relevant question may well be a very speculative one; when looking back 50 years from now, will we see the early twenty-first century as the beginning of a revolution in medicine brought about by the nanosciences? Will the introduction of nanomaterials like silver nanoparticles or the use of liposomes or albumin for vectorization be interpreted as a radical break with the old approach to pharmacy, or as a step on an ongoing path of drug development? Are we living through a new revolution comparable to the chemotherapeutic revolution we presented earlier? For this, we need to look beyond the pragmatic issue of whether we or our children or our children's children will live longer healthier lives thanks to nanopharmaceuticals, and pose the more abstract question of whether nanomaterials and the nano approach to pharmacy will change our conception of medical drugs.

Our unsatisfactory answer to this question is that it is difficult to say. Research is moving very fast, with new approaches mobilizing nanoparticles in novel ways. Fifteen years ago, the vectorization developments in nanopharmacy that we have been looking at were presented as the way ahead, and today researchers are looking more and more toward integrative systems that mobilize several different elements at once. Theranostics has become a new buzzword in nanomedicine, and this will be the subject of the next chapter.

2.16 PHARMACY IS NOT ALWAYS WHAT IT SEEMS

One last warning before passing on to the theme of theranostics: the reader needs to keep in mind that many other concerns apart from purely medical ones play a significant role in the current situation of pharmaceutical research, and this applies as much to nanopharmacy as it does to any other field of research in modern healthcare. The pharmaceutical industry has itself undergone some profound shifts over the last 40 years. First, a series of mergers in the closing decades of the twentieth century reduced the core of the pharmaceutical industry to a dozen or so major players. The businesses that came out of these mergers, like Novartis or Sanofi-Aventis, are now very large multi-national companies that operate on a worldwide scale with increasing investment in marketing and a growing responsibility with respect to the financial demands of investors. One of the results of the re-structuring of the pharmaceutical industry, combined with the importance of genetics and other cutting-edge scientific fields in medicine, has been the outsourcing of innovative research to a collection of high-tech start-ups, often spin-offs from university research. Many of the products we have been discussing (nanovectors, the use of nanoparticles as antibacterial agents) were developed in this constellation of start-ups orbiting around major pharmaceutical companies, and in this context the results from or even rumors about new medicines and their development and testing can make or break fortunes overnight. Thus, the future of a company depends as much on what is communicated about its research as it does on the results of clinical or preclinical tests, undermining the reliability of published information. Over time, the real potential of new products will become clear, but we need to remain cautious in our assessment of innovations based on published discourses about them.

Looking beyond communication addressed to the public, our impression is that researchers and investors are no longer as excited about the vectorization approach to pharmacy as they once were and are now looking to new horizons for innovation in nanomedicine. Theranostic agents and systems constitute one such flourishing area of research, and this approach is the subject of our next chapter.

2.17 TO GO FURTHER

To get an idea of the presence of nanoparticles in medical research, see Anselmo and Mitragotri (2019) and for a general overview of the medical uses of nanoparticles see Rihn (2020), Vasile (2018) and Torchilin (2020). Mertens (2019) looks at recent developments in the lock and key analogy, while Webster (2008) considers the safety of nanoparticles. For up to date analysis of innovation, see Collet (2020) and for more on Liposomes see Bulbake et al. (2017). There are many articles treating stealth design in nanoparticles (Fam et al., 2020 and Li et al., 2020, with Ryals et al., 2020 looking more specifically at PEGylation). Wang et al. (2022) looks at self-assembling molecules.

Anselmo, A. C., & Mitragotri, S. (2019). Nanoparticles in the clinic: An update. *Bioengineering & Translational Medicine, 4*(3), e10143. https://doi.org/10.1002/btm2.10143

Bulbake, U., Doppalapudi, S., Kommineni, N., & Khan, W. (2017). Liposomal formulations in clinical use: An updated review. *Pharmaceutics, 9*(2), E12. https://doi.org/10.3390/pharmaceutics9020012

Collet, F. (2020). *Santé 2030: Une analyse prospective de l'innovation en santé*. Les entreprises du médicament (LEEM).

Fam, S. Y., Chee, C. F., Yong, C. Y., Ho, K. L., Mariatulqabtiah, A. R., & Tan, W. S. (2020). Stealth coating of nanoparticles in drug-delivery systems. *Nanomaterials* (Basel, Switzerland), *10*(4), E787. https://doi.org/10.3390/nano10040787

Li, X., Salzano, G., Qiu, J., Menard, M., Berg, K., Theodossiou, T., Ladavière, C., & Gref, R. (2020). Drug-loaded lipid-coated hybrid organic-inorganic "stealth" nanoparticles for cancer therapy. *Frontiers in Bioengineering and Biotechnology, 8*, 1027. https://doi.org/10.3389/fbioe.2020.01027

Mertens, R. (2019*). The construction of analogy-based research programs: The lock-and-key analogy in 20th century biochemistry*. Transcript Publishing.

Rihn, B. (Ed.). (2020). *Biomedical application of nanoparticles*. Taylor & Francis.

Ryals, R. C., Patel, S., Acosta, C., McKinney, M., Pennesi, M. E., & Sahay, G. (2020). The effects of PEGylation on LNP based mRNA delivery to the eye. *PloS One, 15*(10), e0241006. https://doi.org/10.1371/journal.pone.024100 6

Torchilin, V. (Ed.). (2020). *Handbook of materials for nanomedicine: Metal-based and other nanomaterials*. CRC Press.

Vasile, C. (Ed.). (2018). *Polymeric nanomaterials in nanotherapeutics*. Elsevier.

Wang, H., Monroe, M., Leslie, F., Flexner, C., & Cui, H. (2022). Supramolecular nanomedicines through rational design of self-assembling prodrugs. *Trends in Pharmacological Sciences, 43*(6), 510–521. https://doi.org/10.1016/j.tips.2022.03.003

Webster, T. J. (Ed.). (2008). *Safety of nanoparticles: From manufacturing to medical applications*. Springer.

2.18 REFERENCES

Berube, D. M. (2006). *Nano-hype: The truth behind the nanotechnology buzz*. Prometheus Books.

Crichton, M. (2002). *Prey*. HarperCollins.

Hassiba, A. J., El Zowalaty, M. E., Webster, T. J., Abdullah, A. M., Nasrallah, G. K., Khalil, K. A., Luyt, A. S., & Elzatahry, A. A. (2017). Synthesis, characterization, and antimicrobial properties of novel double layer nanocomposite electrospun fibers for wound dressing applications. *International Journal of Nanomedicine, 12*, 2205–2213. https://doi.org/10.2147/IJN.S123417

Larsen, M. T., Kuhlmann, M., Hvam, M. L., & Howard, K. A. (2016). Albumin-based drug delivery: Harnessing nature to cure disease. *Molecular and Cellular Therapies, 4*, 3. https://doi.org/10.1186/s40591-016-0048-8

Oftedal, G. (2019). The role of "missile" and "targeting" metaphors in nanomedicine. *Philosophia Scientiae, 23*(1), 39–55.

Simon, J. (1999). Naming and toxicity: A history of strychnine. *Studies in History and Philosophy of the Biological and Medical Sciences, 30C*(4), 505–525.

Vandchali, N. R., Moadab, F., Taghizadeh, E., Tajbakhsh, A., & Gheibihayat, S. M. (2021). CD47 Functionalization of nanoparticles as a poly(ethylene glycol) alternative: A novel approach to improve drug delivery. *Current Drug Targets, 22*(15), 1750–1759. https://doi.org/10.2174/1389450122666210204203514

Whitehouse, M. W. (2015). Silver pharmacology: Past, present and questions for the future. *Progress in Drug Research. Fortschritte Der Arzneimittelforschung. Progres Des Recherches Pharmaceutiques, 70*, 237–273. https://doi.org/10.1007/978-3-0348-0927-6_7

Yuan, S., Huang, J., Jiang, X., Huang, Y., Zhu, X., & Cai, Z. (2022). Environmental fate and toxicity of sunscreen-derived inorganic ultraviolet filters in aquatic environments: A review. *Nanomaterials* (Basel, Switzerland), *12*(4), 699. https://doi.org/10.3390/nano12040699

3 Theranostics
Toward a New Integrative Horizon

In the last chapter, we saw how nanoscience has the potential to transform both the design and the production of medical drugs. But the dream of a bottom-up approach to drug development where the active principles are conceived to fulfill precise molecular functions before being made-to-measure, atom by atom, for the moment remains just that: a dream. Thanks to the development of the encapsulation techniques discussed in the last chapter, the nanoparticles that have taken the lead in pharmaceutical medicine have the function of vectors for well-established drugs. As we argued at the end of the chapter, more than 20 years after the approval of the first of these nanovectors for clinical use in the treatment of cancer, it remains hard to evaluate the extent to which they will change the prospects for the long-term survival of someone diagnosed with cancer today. But even before nanoscience has had the time to transform clinical pharmacy through these techniques of vectorization, scientists and the investors who fund them—both public and private—are already looking ahead to the longer term future and diversifying their approaches. Treating disease is no longer an exclusive goal for nanomedicine. Researchers are developing diagnostic and even prognostic tools to facilitate the accurate identification and localization of cancers and other diseases and to help doctors decide on the most appropriate treatment for each patient. Scientists have started to think about all these elements—diagnosis, prognosis and treatment—in parallel in order both to improve existing treatments and to rethink the conception of targeted drugs.

This idea of combining diagnostic techniques—particularly those based on genetics—with targeted treatments has been tied into a wider movement in contemporary healthcare known as precision medicine. While, conceptually, nanoscience is not an essential component of this precision medicine movement, it often features in presentations of the field, especially when the discussion turns to theranostics. As this chapter is about the nanosciences and theranostics, we will take this opportunity to explore the links between the nanosciences and the precision medicine movement.

The field of nanotheranostics has, in part at least, built on the developments in nanopharmacy that we were discussing in the previous chapter, but it lags behind in terms of its clinical applications, so relatively few of the innovations we will be considering in this chapter are available for clinical use. There is a profusion of preclinical research (still confined to biochemistry and medical laboratories), but it remains difficult if not impossible to predict which approaches are likely to translate into viable, let alone useful clinical procedures. Thus, while some of these techniques will doubtless integrate future clinical practice, it is very hard to say which ones,

DOI: 10.1201/9781003367833-3

and even harder to predict what the effects will be in terms of twenty-first-century healthcare.

In this chapter, we will outline this expanding field of theranostics and, with the help of a few examples, we will explain how the nanosciences and nanotechnologies fit into this approach. Before going any further, we need to have an idea of what the term theranostics means, or rather what aspects of research it covers. If you imagine that this is a straightforward question of providing a definition and seeing what research fits this definition, you will have to think again. Indeed, although the origin of the term theranostics is well established, its precise scope has never been entirely clear, and, to make things more confusing, it has developed over the last decades to cover an increasingly large number of approaches in applied medical science. As if these problems in engaging the field were not enough, we need to add the observation that many of the projects qualified as theranostic today fail to match up to even the most basic definition of the term. As a field, philosophy puts a lot of emphasis on getting the definitions of terms clear, and philosophers tend to be critical of those who fail to respect these definitions. Nevertheless, in the domain of theranostics, as in many other active areas of scientific research, it may be just as wise not to be excessively rigorous when it comes to the application of a precise definition, but rather to look at what research is being carried out under the head of the term. This is important if we hope to pick up on interesting innovations, as the most radically transformative approaches in an area like this might well not answer strictly to the definition. Having laid out this methodological warning, however, we can still propose a definition of the field.

3.1 THERANOSTICS—A BROAD DEFINITION

The word 'theranostics' was coined at the end of the twentieth century by combining the two terms therapy (or therapeutics) and diagnosis, and so, in principle at least, this field covers a group of techniques that associate disease treatment with its diagnosis (Wiesing, 2019). The diagnostic side of the definition is already wide ranging; in its most literal sense, it suggests the diagnosis of a particular disease, but it can also be aimed at identifying the appropriate therapeutic approach (drugs, dosage and frequency) or predicting the efficacy of any given strategy for an individual patient with her or his particular genotype. But before going any further, it may well be worthwhile to take a step back from theranostics to think about diagnosis, as it is a term that is used all the time, and yet has not often been the subject of any profound philosophical discussion. In sum, diagnosis has interested philosophers less than therapy and certainly much less than the foundational pair of terms in the philosophy of medicine: health and disease.

A simple and concise definition of medical diagnosis is the use of symptoms* to identify a disease or syndrome*. This kind of traditional semiological vision of diagnosis transports us back to a time when doctors combined an anamnesis* (through skilled interrogation, the doctor has the patient recount her or his problems including the symptoms) with a physical examination to pose their diagnosis, although today in the majority of cases, this description of diagnosis is largely insufficient, with the doctor's physical examination being backed up by a series of laboratory tests. Although the means might have evolved, however, the outcome remains the same;

the doctor's diagnosis puts a name on the patient's disease or syndrome, allows the identification of the cause of the symptoms and, in the best-case scenario, the prescription of an effective therapy.

In a modern medical system, this model of a one-on-one interaction between a doctor and a patient to arrive at a diagnosis only applies in rare cases, with specialized laboratories carrying out a whole range of supplementary tests intended to help the doctor determine the right diagnosis. These laboratory tests serve as much to eliminate possible candidates for the disease diagnosis as to confirm the veritable nature of the pathology.

So how might theranostics change this vision of diagnosis? Potentially quite radically: once a specific diagnostic tool is integrated with an equally specific treatment, we could, in principle at least, dispense with any other diagnostic step. While it would be presumptuous to think that nanotheranostics are going to replace the human doctor in terms of diagnosis, a theranostic substance might be able to use a specific function to carry out an analysis that can serve as the functional equivalent of one or several of the laboratory tests presented earlier. In other words, once a theranostic technique is functioning at a high enough level, there should no longer be any need for the traditional diagnostic step. Doubtless, in the future, artificial intelligence will take on many of the traditional tasks of disease diagnosis, but this subject of diagnosis by artificial intelligence takes us well beyond the scope of this book.

The meaning of diagnosis in theranostics has spread even wider than this initial sense of the term, and now covers any technique aimed at enhancing the effectiveness of a treatment. Thus, most of the examples we will be presenting in this chapter do not involve any diagnosis at all, at least not in the traditional sense of the term. Indeed, today the emphasis seems to be on the contribution of nanoparticles to enhancing medical imaging techniques, such as X-ray, magnetic resonance imaging* (MRI) and positron emission tomography* (PET), as well as the manipulation of customized nanoparticles inside the body from the outside by means of their magnetic or spectroscopic properties in order to increase the amount of a drug that is delivered to a targeted area or in order to activate a therapy from without the body.

Reflecting such an inclusive definition, some of the oldest projected uses of the nanosciences in medicine now find themselves classified under the head of theranostics, although they do not necessarily involve any diagnostic techniques in the more traditional sense of diagnosis. Indeed, forms of drug delivery that offer the possibility of enhancing the targeting process or simply increasing the visibility of a treatment target or the therapeutic action of a drug using modern imaging techniques are now considered part of this domain. If we take as our starting point what today's scientists consider to be theranostic research or projects, the working definition seems to be that of any system that mobilizes certain substances in order to deploy disease treatment as a follow-up to another function (the enhancement of imaging or the localization of a tumor, to give just two common examples) using an integrative approach. Nevertheless, as we have already remarked, this integration does not necessarily provide a diagnosis in its traditional sense.

In light of these initial observations on theranostics, we can already see that the domain is not easy to circumscribe. Thus, in spite of the fact that we just gave the definition, we are going to need to present several examples of the kinds of project that are today classed as theranostic in order to make the (practical or operational) sense of the term clear. Our plan is to start by explaining the principle of theranostics around a well-established case of research drawn from oncology, the use of hormone receptor status to modulate the treatment of breast cancer, but it is a technique that does not integrate the nanosciences, at least not for the moment. We will then go on to present a series of examples of theranostic techniques that do rely directly on the nanosciences. The last part of the chapter will return to the question of diagnosis in theranostics and will lay out the links between this approach and the rising field of precision medicine. Thus, prior to presenting a selection of examples of contemporary theranostic projects that involve nanoparticles, we would like to clarify the principles behind combining diagnosis, particularly the detailed diagnoses arising out of genetic analyses, with therapy, and we will do this by using the very rich example provided by the diagnosis and differential treatment of breast cancer.

3.2 BREAST CANCER: APPROPRIATE TREATMENTS FOR DIFFERENT TUMORS

A patient who discovers a suspicious lump in their breast[1] will routinely undergo a biopsy: the surgical removal of a small sample of this growth, which is then sent to a specialized pathology laboratory, primarily to determine whether or not the lump is a malignant cancerous tumor. Anyone who has had this experience or knows someone in this situation will be aware that the report that comes back from the laboratory is not limited to an assessment of the general status of the tumor: cancerous or benign. The analyses of these biopsies provide a great deal of additional information, with different laboratory tests, notably genetic tests, being used to classify or qualify the tumor. One test that is routinely carried out is to establish the presence of receptors for certain hormones in the tumor, in particular the receptors for the two hormones estrogen and progesterone. Besides telling you whether the tumor tests positive or negative for the presence of these receptors, the report will generally give you the percentage of cells found to express them.

Both of these hormones, estrogen and progesterone, are present in everyone's bloodstream, and in ovulating women their concentrations change across the menstrual cycle. Indeed, both hormones were used in the first birth control pills dating back to the 1950s with the aim of disrupting this cycle and thereby preventing pregnancy (Oudshoorn, 1994). Looking at just one of these elements of the biopsy report, the estrogen receptor status—positive if the tumor expresses these receptors or negative if it does not—we can see that it has become an important parameter in deciding what treatment options to pursue and is also a key factor in predicting the evolution and ultimately the outcome of the disease. Even in the absence of details about the different mechanisms at work in these cancers, the long history of these tests and the correlation between the results of the test and the results of the different treatments that followed has provided a lot of useful empirical data. Statistically significant differences in outcomes for the same treatment of tumors given to patients with different hormone receptor status provide invaluable help for the medical team

[1] The reader should bear in mind that around 1 percent of breast cancers occur in men.

in choosing the appropriate treatment for their patient. Globally, being hormone receptor positive is a positive prognostic indication (in particular, the patient is more likely to respond well to chemotherapy), although it is associated with higher rates of relapse. Furthermore, if someone is hormone receptor positive (for estrogen, progesterone or both) doctors can prescribe complementary treatments to artificially lower the levels of these hormones in the body, a precautionary measure that is unnecessary for patients testing negative for these hormone receptors. Setting aside these details for the moment, the central idea is that a specialist laboratory can perform a test on the material obtained from the biopsy to establish the hormone receptor status of a tumor, and the results of these tests have an influence on the treatment that will be proposed by the doctors in charge of the case. The hormone receptor status that is added to the diagnosis of breast cancer has an influence on the choice of therapy. We are in the realm of theranostics; a diagnostic technique has a direct influence on the therapeutic protocol put into place. As we will see later, we are also here on the frontiers of precision medicine, where the treatments that the doctors prescribe are tailored to the characteristics of the tumor and the patient taken in charge.

We can go still further with this breast cancer example, as hormone receptor status is not the only parameter that is established by the analysis of the biopsy performed in a pathology laboratory. Another element that the laboratory will test for is the presence of the human epidermal growth factor receptor (HER 2) (Harbeck et al., 2022). The overexpression of this HER-2 receptor in the tumor will generally result in treatment with the drug trastuzumab (Herceptin® for the Roche product), usually in combination with other drugs, at least in the populations that are able to pay for this treatment. Trastuzumab will not be proposed to a patient who tests negative for the HER-2 receptor, as the drug specifically targets this receptor.

Thus, as an example, Herceptin® is a significant drug in several respects. A monoclonal antibody*, it was developed and marketed at the end of the twentieth century by Genentech, one of the leading companies in the wave of genetics start-up companies dating from the 1970s. But of more importance to us here, it is a drug that was linked from the early stages of its development to a test for a specific protein. Herceptin is, in principle, used only in patients who express the HER-2 receptor, making it one of the first drugs to apply the theranostic principle of combining diagnosis with therapy or, more accurately, conditioning a particular treatment on the result of an immunohistochemical test.

We have started with an example that is at the limit of the field of theranostics, as while the hormone receptor approach to breast cancer involves both diagnosis and therapy, the two are separated in time and in terms of the technologies (immunohistochemical test followed by treatment with Herceptin®) that are put into place for the diagnosis and the therapy. Furthermore, in this particular case, the nanosciences have not been directly mobilized in either aspect of this technique: diagnosis or therapy. Nevertheless, it is not a great conceptual leap to envisage appropriately designed nanoparticles to test specifically for the presence of HER-2 receptors. A molecule that binds to this receptor and also possesses a fluorescent or radioemitter functionality to signal the presence of the receptors in the tumor or even elsewhere, when cancerous cells are disseminated in the body. If this specifically binding molecule (generally called a ligand*) could be used to deliver a treatment that targets cells that overexpress this receptor as well, the combination of the two; diagnosis—or in this case detection of the presence of the receptor—and treatment would provide an excellent illustration of theranostics in the domain of the nano.

The laboratory analysis of tumors and their hosts does not end with HER-2, and researchers are constantly in search of genetic mutations or receptors whose presence will make a difference to the prognosis of a cancer, contribute to the decision about the appropriate treatment, or offer a novel target for drug therapy. We will return to this issue when we consider the wider field of precision medicine, as these kinds of consideration are at the heart of this approach. Meanwhile, to make the link between theranostics and nanomedicine, we will now look at a few examples of theranostic techniques involving nanoparticles. We want to present several of these systems in order to give the reader a feel for the range of techniques that fall under this heading of theranostics. Thus, the five projects outlined in this chapter should provide an idea of the kinds of directions this research is taking, although they are far from being exhaustive. Like all branches of the nanosciences, theranostics has grown rapidly and diversified in recent decades, and when the dust begins to settle (if it ever does, of course) and it becomes clearer which areas of research are the most promising for further development, it would not be surprising to see theranostics further sub-divided into more delimited specialties. For now, however, theranostics remains a well-identified area of clinical and fundamental research with its own journals and scientific conferences, two signs of the health of a scientific discipline.

3.3 RADIOLIGAND THERAPY—THE DUAL FUNCTIONALITY OF RADIOACTIVITY

A radionuclide is a radioactive element, meaning that the substance made up of this element is unstable and its atoms can break down at any moment to give those of another element: a decomposition at the atomic level that is accompanied by the production of ionizing radiation. The rate at which one of these radioactive elements breaks down is measured by its half-life, which is the length of time it will take for half of a sample of the element to undergo radioactive decay*. A shorter half-life generally means that the radioactivity is more intense, but the rate of decay drops off more quickly. Gallium-68 (Ga68), for example, is a radioactive isotope of the element gallium with a half-life of just over an hour. Ga68 breaks down to give the element zinc accompanied by the release of beta radiation, composed of positively charged sub-atomic particles called positrons. This property of generating ionizing radiation is the basis for the dual use of these elements in ligands as a theranostic approach involving the localization and treatment of cancerous cells.

A ligand is a molecule that, because of its chemical structure, binds very specifically to certain other molecules with a high affinity. To make a change from the famous lock and key metaphor that is standardly used to describe this specificity, we can use a hand and glove metaphor, with the ligand likened to a finely tailored glove that only fits on to one particular shape of a hand, that of the targeted receptor. A radioligand* is simply one of these molecules with a radionuclide inserted somewhere into its structure, giving a radioactive version of the ligand. These substances can be put to one of two uses depending on the profile of the radioactivity they generate. Thus, one profile of emission can be used to enhance medical imaging, helping radiologists to identify and localize tumors in the body; gallium-68 inserted into a ligand that attaches to a receptor known to be found predominantly in cancer cells can indicate the presence of these cells through its radiation signal. The emitted radiation is picked up by imaging systems like PET scans, allowing the doctors to see where the cancerous cells are located and to evaluate the involvement of different organs around the body (Barbosa et al., 2022). This is particularly useful when the tumors become disseminated throughout the patient's body, a phenomenon called metastasis, which designates the process of the spreading of a cancer from an initially localized tumor. This being said, with certain cancers, the disease can be spread throughout the body from the outset.

Radioligands are already in use to help localize neuroendocrine tumors, for example, which are cancerous growths that arise in the body's hormonal system (the endocrine system) or its nervous system and are rarely localized in a single area. By placing gallium-68 in a ligand that binds specifically to a receptor known to be overexpressed in these cancer cells, called the somatostatin receptor, oncologists can localize the somatostatin-positive neuroendocrine cancer cells in the body. For this, they introduce the nanoparticle and then use a PET scan or MRI to localize the areas where the molecule is found in high concentrations. As the ligand attaches preferentially and vigorously to the somatostatin receptors, the visibility of the ligands translates into the localization of the cancer, and this is the first step in the strategy for improving the treatment of the disease. For an example of work with meningioma, see Roytman et al. (2021). Different radionuclides or radioisotopes inserted into the same ligand can then be used to treat the cancer directly by radiotherapy. The radiation released by the decaying element is no longer just used to localize the tumor through PET imaging but now serves to kill the cancerous cells, thanks to the damage caused to the cell's DNA. In general, this radiation induces the production of reactive oxygen species within the cells that are responsible for provoking DNA damage; with the cell's mismatch repair mechanisms unable to cope, a sequence of internal signals leads to apoptosis. Thus, these somatostatin-specific radioligands are theranostic nanoparticles, with the same molecule performing both roles, the diagnosis of the disease and then its treatment.

This being said, today the diagnosis of a neuroendocrine cancer would have to precede the use of the radioligand for either of the purposes described earlier; location by imaging or treatment by radiation. In sum, the use of radioactivity to locate and visualize the tumors in the body is not, strictly speaking, a technique used for the diagnosis of the disease, but rather a refinement or an enhancement of the detection process to improve the medical team's knowledge of the distribution of the cancer in the patient's body. Still, it is clear that this kind of technique could, in principle at least, be used to make or confirm an initial diagnosis of cancer.

3.4 REMOTE GUIDANCE AND IMAGING—THE THERANOSTIC POTENTIAL OF MAGNETIC NANOPARTICLES

The radioligand approach that we just looked at mobilizes radioactivity in two different ways to offer an integrated imaging and treatment approach to certain cancers.

Magnetism is another property of matter that can serve different functions and so has attracted the attention of researchers for its potential use in theranostic applications.

At first glance, however, the use of magnetism to treat cancer resonates more with a long tradition of alternative medicine than with cutting-edge nanomedicine. Although they enjoyed their heyday at the beginning of the twentieth century considered at that time as hi-tech, mysterious electromagnetic apparatuses are still touted as cures for anything from headaches to chronic backache. These machines are more likely to be sold on cable television networks than in a prescription pharmaceutical outlet and are widely dismissed by those who work in mainstream medicine. Most anti-cancer drugs, like the doxorubicin that we looked at in the last chapter, work by interfering directly with the cellular mechanisms that allow cancers to grow out of control. In order to disrupt cellular function, the drugs have to be introduced into the cells; it is not the kind of treatment that can be applied at a distance. Magnetism, on the other hand, is a force that acts at a distance, and, unlike the ionizing radiation we discussed earlier, it only affects magnetic materials and appears to pass through the intervening non-magnetic matter without any noticeable effect. While the majority of the body is made up of non-magnetic material, magnetic fields affect electrical systems, including the central nervous system. Thus, the health effects of magnetic fields on humans is a controversial topic, and with the constant augmentation of electromagnetic radiation in the environment (high-voltage electrical power lines, radio waves, television, and now Wi-Fi and cellphone signals), these controversies are liable to gain ground in the future. Unlike ionizing radiation, magnetism is not currently used to treat cancers directly, or at least not in the context of mainstream medicine, as we suggested earlier, meaning that the theranostic use of magnetic nanoparticles might seem a less obvious option than the use of radionuclides, as radiotherapy (using radiation to cure cancer) has been a conventional treatment in oncology for more than a century now. On the other hand, the magnetic properties of matter have been central to the development of the nanosciences, as they offer a number of novel opportunities at the nano level. Because magnetism depends on the movement and orientation of sub-atomic particles, especially electrons, nanomaterials can produce very interesting magnetic effects. Light, for example, can be used to induce magnetization, an effect that is not seen at the scale of macroscopic objects.

While such electro-magnetic relationships and quantum properties of matter may present opportunities for ground-breaking therapy in the future, for the moment the use of magnetism in nanomedicine is less ambitious, mobilizing the possibility of guiding magnetized matter from a distance and the potential for rapidly alternating magnetic fields to generate heat in sensitive magnetized nanoparticles, a property which can be used to damage or destroy cancer cells in the vicinity of these particles (Theodosiou et al., 2022).

The most easily observable magnetic property that we come across in our everyday lives is the one associated with iron and other rarer metals like cobalt and nickel, and this is called ferromagnetism. It is this ferromagnetism that explains why your fridge magnets stay attached to the stainless steel door of the fridge, as there is a strong force of attraction between the iron in the door and the magnet. But these same magnets will fall off an aluminum can. Aluminum, like magnesium and sodium, is a paramagnetic material meaning it is only weakly attracted by a magnetic field.

Iron oxide is a magnetic compound that has been produced in the form of nanoparticles and has already been widely used in the experimental application of magnetism in nanomedicine. More specifically, superparamagnetic iron oxide nanoparticles, also known as SPIONs*, are a family of magnetic nanoparticles that are already being used in experimental medical research both in vitro and in vivo. SPION-based materials have been approved for use as contrast agents to improve magnetic resonance imaging* (MRI), and, as we have seen with the logic of theranostics, this function can serve as the basis for exploring other complementary uses, notably the use of magnetism in disease treatment. Chemists have been able to place SPIONs on the surface of liposomes like those we were looking at in the last chapter, and the presence of these magnetic particles thereby enables the external guidance of drug delivery; magnets, either temporary electromagnets or permanent magnets, positioned outside the body can be used to guide liposomes loaded with an anti-cancer drug toward a tumor thanks to these SPIONs or another magnetic element attached to the vector. This kind of guidance combined with the enhancement of MRI imagery brings us into the realm of theranostics, but the initiatives around magnetism do not end here. Nanoparticles featuring an iron oxide core allowing magnetic attachments such as SPIONs can be used to damage or destroy cancerous cells directly. Exposing magnetic particles to a fluctuating magnetic field causes them to heat up, and there are several preliminary studies that are looking into using this property to kill cancerous cells directly (Pucci et al., 2022).

This kind of thermal therapy (magnetic hyperthermia) can be combined with the other uses of these magnetic nanoparticles that we have just presented. With an iron oxide core, a nanoparticle could enhance MRI, and by heating it up using an alternating magnetic field, it could contribute directly to an anti-tumor therapy. Adding specific ligands to the surface of the particle would allow its preferential combination to a receptor associated with a targeted cancer, increasing the precision of both the imaging and the thermal therapy. So SPIONs offer another example of a multi-functional nanoparticle that can contribute to integrated approaches combining therapy with imaging. Although these approaches are classified under the head of theranostics, once again, it is not clear that any diagnosis is being made. But we will return to this point, as it is something that many of these theranostic approaches have in common.

3.5 THE PROSTATE-SPECIFIC ANTIGEN (PSA) AND PROSTATE CANCER

Our third example of nanotheranostics takes us back to the nanoparticles as drug vectors that we were looking at in the last chapter. We already saw that starting with the third generation of these vectors, chemists have been able to 'customize' the liposomes, attaching various molecules to the vector, including ligands for known receptors. These ligands bind specifically to cells bearing these structures (we have already seen that this is the key property of the family of chemicals known as ligands), enabling the targeted delivery of an anti-cancer agent for example. Starting from this premise, once a receptor has been associated with a particular type of cancer then there is, in principle at least, a way of delivering an anti-cancer agent directly and predominantly to the cancerous cells, even when they are disseminated throughout the body. We can see a generalized therapeutic strategy taking shape (on the basis of selective uptake via a receptor), and the treatment of prostate cancer, particularly when it has left its primary site (the prostate) is one disease in which this strategy is being actively pursued.

Since the end of the twentieth century, the prostate-specific antigen* (PSA) has become the benchmark indicator for the presence of prostate cancer. In older men, even those without any symptoms, the concentration of the PSA in the blood gives the first indication that someone might have prostate cancer, whether symptomatic or not. Although the concentration of this molecule varies between individuals as well as cyclically in the body, it remains the first signal detectable by laboratory analysis that a man may have this cancer. The level of PSA has become a standard element in blood tests, and an elevated level of this antigen will be signaled on the results of most blood tests and elevated concentration will normally lead to further investigation of the state of the patient's prostate gland. Nevertheless, while the level of PSA can trigger suspicions, the diagnosis of the disease is generally done through medical imaging or palpation of the gland before being confirmed by a biopsy.

If the prostate cancer spreads to other parts of the body in a process known as metastasis—a development that does not augur well for the patient—the cancerous cells continue to express the PSA, an antigen associated with the primary prostate cancer, even though they may now be located far from the patient's prostate. As with the radioligands we considered earlier, a liposome customized with a PSA-specific ligand will preferentially attach to the sites of the cancer, wherever it might be in the body. Loading the liposome with nanoparticles that can enhance imaging (gold nanoparticles, for example, or radionuclides as in our first theranostic example) helps to localize the sites of the disease, and combining this with an anti-cancer agent like docetaxel can lead to the directly targeted treatment of the disease. While still in the preliminary stages, researchers hold out considerable hope for this kind of theranostic approach where the same vectorized molecule can simultaneously allow doctors to localize tumors with the aid of PET or other imagery, and provide them with the means to treat the disease directly. We could describe this strategy as killing two birds with one stone, as the patient would only have to receive one dose of the appropriate multi-functional molecule, reducing an intrusive intervention to a minimum. Nevertheless, looking deeper into the treatment of prostate cancer, we can see that the approach also raises important questions.

Within urology, there has been a movement of doctors expressing concern about the possible over-treatment of prostate cancer. The argument is that many prostate cancers develop very slowly and because they often occur in older men, the patient is likely to die before the prostate cancer would cause any significant health problems. Treating the cancer, usually by the surgical removal of the gland, can, on the other hand, trigger substantial problems, meaning that for certain patients the detection and resulting treatment of their cancer could have an overall negative impact on their health when compared to the tumor being left untreated. While we are not qualified to take sides on this particular debate, it is interesting to consider the problem and how it relates to theranostic approaches to prostate cancer (and, as a corollary, to other diseases that might be subject to this kind of approach). A multi-functional theranostic agent, like the one described earlier, already has the decision to engage in chemotherapy built into it, without any assessment of the potential risks associated with this form of treatment. If this integrated approach ever became widely adopted for confirming and treating suspected cases of cancer, then specialists would have to rethink the decision-making process. The use of this kind of theranostic agent could be reserved for patients who had already undergone an independent diagnosis, rather than delegating the decision to treat or not to the theranostic system itself.[2]

3.6 TRANSFERRIN AND BIOLOGICAL BARRIERS. ANOTHER FORM OF THERANOSTICS?

The approaches involving the transferrin* receptors both take us away from the exclusive focus on cancers and introduce an example of how to turn a molecule with interesting biological properties to the ends of nanomedicine. Whether it merits the qualification of theranostic remains to be demonstrated, but the use of transferrin is often cited in lists of theranostic innovations. Thus we feel warranted in including it among these examples. Transferrin is a protein that binds to iron in plasma and transports it to cells. In the process, transferrin attaches to transferrin receptors, triggering a mechanism that allows the protein to overcome formidable biological hurdles such as the blood-brain barrier, a biological frontier that protects sensitive neuronal tissue from potentially dangerous molecules present in the blood. Unfortunately, this same blood-brain barrier makes it difficult to use drugs to treat conditions that affect the brain and central nervous system, such as Parkinson's disease or Alzheimer's disease. With aging populations around the world, these diseases are becoming more common, making the need for effective treatments ever more pressing. Thus, researchers are particularly interested in finding ways to deliver drug treatments to the brain and the rest of the central nervous system more effectively, and mobilizing the transferrin molecule as a means to transport molecules across this barrier offers a chemical basis for nanoscientists to do just that. By modifying transferrin with cytotoxic compounds or other therapeutic agents, scientists can exploit this capacity to cross the blood-brain barrier to reach targets in the brain (Wang & Tang et al., 2022). Using transferrin to 'piggyback' a drug into the brain seems to have taken us a long way from the theranostic ideal of combining therapy with diagnosis, but, as we have already noted, this kind of approach is often included in lists of theranostic innovations, so it merits a place in this section, reminding us of just how far the notion of theranostics has already spread from its initial meaning. At the risk of remaining in this category of techniques that do not clearly fit the initial definition of theranostics, we want to conclude with an example illustrating a very different type of approach, bringing nanoparticles into partnership with surgery rather than with chemo or thermal therapy.

---

[2] For a more complete discussion of this issue, see Simon, 2021.

3.7 VISUALIZING FOLATE RECEPTORS DURING SURGERY

In most cancers where it is possible, surgical removal of an initial tumor is a more effective treatment than chemotherapy alone, notably in terms of long-term survival. This being said, both surgery and chemotherapy are regularly used together, often combined with radiotherapy to maximize the patient's chances of survival. To be effective, the initial surgery has to be thorough, as leaving clusters of cancerous cells behind after the intervention increases the risk of the cancer re-occurring or, worse still, metastasizing*. It can be very difficult for the surgeon to distinguish cancerous from healthy tissue in the vicinity of an operable tumor, so any aid for visualizing cancerous cells is of considerable value in the operating theater. Folic acid is a molecule essential for the functioning of cells, and when it arrives at the surface of a cell, it binds to a folate receptor before being drawn into it for use. These folate receptors are overexpressed in the membranes of a large number of cancer cells, as with ovarian or lung cancer, meaning that the cancerous cells have more of these receptors than the healthy cells around them. Injecting a nanoparticle with a ligand that both binds to the folate receptor and fluoresces after excitation with infra-red light allows surgeons to visualize concentrations of cancer cells during the operation. In real time, the clusters of cancerous cells glow on an infra-red imaging screen, allowing the surgeon to locate the tissue that she or he needs to remove in the vicinity of the principal tumor.

We wanted to bring up this example, because it is quite different from the others. The nanoparticle is not in itself contributing directly to the therapy but acts as an aid to the surgeon, and it is not clear, as with our other examples, to what extent it is really contributing to the diagnosis of this cancer either. Nevertheless, this approach fits with the more general definition of combining functionalities, as the preferential binding property of the nanoparticle is linked to a fluorescent function thereby helping to render the cancer visible. Once again, it is an example that is often cited in discussions of theranostic techniques. We can see that in the broadest sense the specificity provided by the folate receptors is linked to an infra-red visualization functionality, meaning that it fits the broad definition of the theranostic as an integrated multi-function technique. This approach also opens the door to identifying and visualizing many other types of tissue characterized by the overexpression of a known receptor through appropriately designed nanoparticles. If a nanoparticle were able to identify a diseased cell susceptible to respond to a particular type of therapy, it could, in principle, signal this directly to the medical team, allowing the appropriate adjustment of the treatment.

Before exploring the many links that exist between theranostics on the nano scale and the precision medicine movement, we would like to draw out a few conclusions concerning the current orientation of nanotheranostics based on the examples we have just been looking at, starting with the place of diagnosis.

3.8 DIAGNOSIS AND TREATMENT: THE DAWNING OF A NEW ERA?

When we think about modern medicine, we tend to think of treatment and cures, making disease or illness just the starting point for medical practice. People seek medical attention in the first place because they feel ill (illness as the subjective feeling experienced by the patient) and usually assume they feel ill because they are suffering from a disease (the objective element that is the cause of their illness).[3] Although patients no doubt appreciate a doctor putting a name on their medical condition, and it is clearly an important part of the ritual of the medical consultation, the main reason they go to see their doctor in the first place is because they want to get better, be cured, or at least return to an acceptable state of health, if not what they consider normal. This being said, in the classic structure of a medical consultation, the diagnosis precedes the treatment (and, ideally the cure) and it is the diagnosis that orients, if not determines, the appropriate treatment. Diagnosis is important, and misdiagnosis is a major problem in modern medicine. Just as an example, we can recall the time and effort (and considerable government investment) that have been spent on trying to improve the differential diagnosis between bacterial and viral infections in order to reduce the inappropriate use of antibiotics to treat viral disease. In the ideal clinical situation, an accurate diagnosis is followed by the prescription of the appropriate treatment (be it drug therapy, surgery, radiotherapy or any other therapy) and the patient is cured.

Even when there is nothing that can be done to cure or even treat a patient, an accurate confirmed diagnosis can be of vital importance. To stop the spread of an infectious disease, for example, it is crucial to know who is and who is not infected, even if no effective treatment is available. This has been made abundantly clear by the recent COVID-19 pandemic. As we have seen on numerous occasions, accurate diagnosis, even without any available treatment for the disease, can be a matter of life and death.

This being said, historically, the discovery of an effective treatment for a disease has changed not only the effect of the diagnosis but also its meaning for patient and physician alike. In the early 1980s, being diagnosed with AIDS* (a result of infection with the human immunodeficiency virus (HIV), although the virus itself was not identified until 1983) was to all intents and purposes a death sentence, with doctors unable to propose anything but symptomatic treatments for the opportunist infections accompanying the underlying disease. Except in a few cases of spontaneous immunity, the disease would inevitably get worse and the patient would die. While the infection with HIV remains incurable today, there are highly effective treatments (on the basis of tritherapy, which dates from the 1990s but is continually improving) that can reduce the virus to a minimal level and allow patients to avoid any associated opportunistic diseases. Thus, being diagnosed as infected with HIV today holds out quite different prospects from those of 35 years ago; a patient receiving the appropriate treatment can expect to have a normal life span, can have HIV-free children, and will not necessarily develop any of the symptoms of AIDS. Refinements in diagnosis (determining the patient's viral load or identifying the strain of the virus) can lead to appropriate adjustments in the treatment to ensure its continued effectiveness. In the last 40 years then, the nature and the significance of a diagnosis of being

[3] For more on the distinction between illness and disease, see Boorse, 1975.

infected with HIV have changed dramatically, largely because of developments in the associated treatment.

Theranostics proposes bringing diagnosis and treatment closer together, and nanotheranostics aims to do this using one single product at the level of the nano that unites the two functions. This approach implies that there is already an effective treatment for the disease in question, and this treatment is integrated into the nanotheranostic particle or device. As we have been trying to suggest, the logic of the development of theranostic techniques is that the diagnosis and treatment can one day become autonomous. Nevertheless, the end of the independent practice of diagnosis prior to treatment remains a long way off. For the moment, that is to say in all the examples we have presented here, a clinical diagnosis would have to precede what are essentially experimental uses of these theranostic techniques.

3.9 THERANOSTICS AND PRECISION MEDICINE: REFINING AND INDIVIDUALIZING TREATMENT

Precision medicine (or personalized medicine, to use a term that was popular until a few years ago) is a relatively new name for an approach or maybe attitude that most would agree has a long pre-history in healthcare. The overall goal of precision medicine is to escape from a one-size-fits-all 'industrial' style that characterizes modern medical care, and refine the healthcare regimen, making sure it is the best adapted version for the individual under treatment. While this image of medicine as the indiscriminate application of rules exaggerates the industrial features of modern medicine, it does reflect a reality of increasingly uniform medical practice (at an international as much as national or even regional level), dictated by consensus protocols and all played out against an institutional background bound by sets of standard operating procedures. Today, once a disease diagnosis has been made—non-small-cell lung cancer, pulmonary tuberculosis, Creutzfeldt-Jakob disease, etc.—the doctor or a larger, often interdisciplinary medical team, will, where possible, put a treatment into place (or no treatment at all). Precision medicine proposes multiplying the laboratory tests, particularly genetic tests performed on the patient, and then selecting the best treatment in light of all the available data concerning previous outcomes with the different treatments available, analyzed as a function of the results of these tests. Thus, the idea is to take as many (pertinent) factors as possible into account for each individual patient and their specific disease when determining what course of treatment to pursue, with genetic and environmental factors heading the list. The result should be a therapy that is finely tuned, as a function of the patient's characteristics and those of the disease: one that maximizes the chances of clinical success. With the price of sequencing a person's entire genome constantly falling, it is easy to imagine that this data will be available to an increasing number of patients seeking treatment for a serious disease.

The slogans around precision medicine are very seductive; it is going to put an end to 'one-size-fits-all' medicine and it is an approach that 'takes into account individual variability in genes, environment, and lifestyle' (https://ghr.nlm.nih.gov/, 31 May 2020). The logic is clear. First, every individual is just that; an individual, with different circumstances and personal histories (genetic, environmental and lifestyle) that have brought them to their present health situation. These patients can also be very different in terms of their reactions to drugs or other medical treatments. Doctors have always known that when the same drug is given to different people suffering from the same disease, no two people will respond in exactly the same way in terms of either the therapeutic effect of the drug or its side effects. The most extreme cases are seen in people who are allergic to a drug, and while they are rare, every year there are deaths due to drug allergies triggering what is known as anaphylactic shock*. Anyone who has been hospitalized knows that in order to avoid these kinds of incidents you are asked to declare any drug allergies on admission, especially allergies to antibiotics like penicillin. Dividing patients into those who are allergic to penicillin and those who are not might be considered a first step down the path of precision medicine, but the project aims to go a lot further than this.

Under this regime of precision, doctors will use tests, principally DNA tests (Napoli et al., 2022), to divide the population of patients up according to their biological constitution and personal histories in order to be able to predict which individuals are likely to respond well to a particular treatment and which ones are not likely to do so. This information in turn comes from processing the data on other patients who have received the treatment before, as well as from knowledge about the functioning of the drug (how fast the drug is metabolized and how this takes place). The data can be treated by artificial intelligence systems to reveal and measure correlations, allowing the doctor to propose the most effective therapy in light of all the available information.

The emergence of precision medicine as a domain over the last 20 years is tied to the rise of big data in medicine and trans-disciplinary initiatives aimed at exploiting this data in order to improve treatment outcomes for each individual patient. The large amounts of data generated by an ever-increasing number of tests and trials of treatments conducted on large cohorts of patients have fueled this project of individualizing therapy patient by patient. Taken to its extreme conclusion, precision medicine would be its own undoing; in a world where every patient received a specifically tailored treatment, no two treatment regimens would be the same, and the comparative data that justifies the individual treatment choice would eventually dry up. In a world where no two patients receive the same treatment, any statistical comparison of outcomes would be impossible.

Like much scientific research, precision medicine is premised on a system of trial and error informed and modified by data feedback. In an illusory ideal, this approach would transcend the error part of the method, and each patient would receive the precisely determined optimal treatment based on the maximal amount of individual data (concerning both the patient and the disease). While intriguing, this paradox is purely theoretical, as with the current size of the human population, the tailoring of treatment is unlikely to reach these extremes of individualization, and so the medical world does not risk running out of data on differential results from the same or similar treatment for the same or similar condition. Indeed, one of the reasons that the expression 'precision medicine' came to be favored over 'personalized medicine' is because the ideal of each patient receiving a different, bespoke treatment as suggested by the latter phrase is simply not realistic, except for a handful of the most privileged (usually the wealthiest) patients.

This being said, doctors have always taken the patient's individual circumstances into account in their clinical work, and adapting a patient's treatment to her needs, capacities and possibilities is an important element of the prescription process. Precision medicine takes this approach by supplementing the doctor's clinical judgment with the results of all that is known about the disease, its development and the likely response of a patient with a specific (genetic and epigenetic) disposition to any given treatment. Here there is an issue that troubles many doctors: is the ultimate goal of this precision approach to supplement the clinician's judgment or to replace it? With artificial intelligence already threatening to diagnose disease in the doctor's place, will precision medicine eventually do the same for the prescription of an appropriate treatment? Here, we can raise another major medical movement that has marked the beginning of the twenty-first century, Evidence-Based Medicine (EBM)*. Put crudely, EBM argues for basing treatment choice on the results of clinical trials rather than relying on the doctor's own personal judgment or beliefs. The principle is relatively uncontroversial in countries like the United States and the United Kingdom, although individual applications of EBM have been criticized everywhere. In other countries, it is not uncommon to hear criticisms of even the principle of EBM, as many doctors believe that the individual doctor's knowledge of their area of specialization combined with their clinical experience leads to better individual judgments about treatment than the blind application of a consensus protocol, what many equate with the application of the principles of EBM.

3.10 GENETICS AND CANCER

For the time being, the majority of the proposed applications of this genetics-based approach to profiling drug treatment have been in cancer chemotherapy, so we will describe the principles behind precision medicine with respect to this area of application. The reader should bear in mind, however, that the same approach could (and already does) apply to any form of treatment for any disease and precision medicine is not solely concerned with cancer.

On the side of genetics, data concerning the correlation between response to a drug and a particular mutation can give rise to a genetic test that will predict whether a patient will respond positively to this drug. This is generally done statistically, with researchers looking for correlations between genetic mutations and the likelihood of the patient having a positive response to a drug that is under consideration for chemotherapy. Once the analysis is done, the drug, which before might have been prescribed to everyone diagnosed with a particular pathology (non-small-cell lung cancer, for example), can be proposed to some and not to others as a function of their predicted response.

Following this brief presentation of precision medicine, we hope that the reader can see why theranostics is often included within its scope. The specificity of theranostic techniques takes into account inter-individual differences. Consider our first example, the use of radioligands to localize and then treat somatostatin-positive neuroendocrine cancers. If radiotherapy of somatostatin-rich cells in this condition can be shown to be preferable to other possible treatments for this specific disease in terms of remission of the cancer or, ideally a proven increase in life expectancy, then the nanotheranostic approach will be validated as a treatment option based on this specificity of the tumor. The multi-functionality of nano-platforms or vectors could enable a specific mode of diagnosis to be integrated with a paired mode of treatment. If the combination of the diagnosis and the treatment is shown through trials to be more effective than the standard global approach to the disease condition, then the theranostic approach has a good chance of being validated as it physically integrates the two. But from the perspective of nanomedicine, one might want to ask what is there to gain from this association with the much broader precision medicine movement? For the moment, the principal answer seems to be access to public funds for research.

In January 2015, Barack Obama inaugurated America's Precision Medicine Initiative (PMI). For the US government, the heavily funded program was as much about keeping America's innovation economy ahead of the competition as it was about improving health outcomes for the US population. In this respect, the PMI neatly paralleled the National Nanotechnology Initiative inaugurated by Bill Clinton 15 years earlier, which broadly boosted all areas of nanoscience and nanotechnology in the United States, and by an effect of imitation if not emulation in other innovation economies, elsewhere as well. Thus, nanotechnology and precision medicine are both areas that were identified, nationally and internationally, as being of high economic potential. So, it is not so much of a coincidence that there is common ground between them in terms of both research and its applications.

Heavily funded as it is, precision medicine (under whatever name) is an important trend in contemporary healthcare that will be shaping the field for decades to come. While it is almost certain to be disappointing with respect to the very ambitious goals (if not promises) that were used to justify such a large initial investment in the project, we could make the same observation for all these kinds of large federally funded projects. Thus, it would be churlish simply to denounce the timeline or the broad, overoptimistic pretensions that informed the PMI's launch. As with other such projects, like the Human Genome Organization project (HUGO), it is not because it is impossible for the project to live up to the initial promises that were made at its inception that we can dismiss it lightly. Precision medicine will, as we have said, be an influential movement in the coming decades. Looking back at the history of biotechnology and genetics, it is all too easy to find exaggerated predictions and promises of what the biotech revolution would bring about, but it is impossible to deny the profound impact that genetic research in general, and the sequencing of DNA in particular has had on human health and agriculture as well as many other areas of our lives.

Just as personalized medicine has today been transformed into precision medicine, maybe in 10 years' time we will no longer be talking of theranostics. Even if it is replaced by another term—'integrative' or 'inter-modal' medicine—the innovations will no doubt be impressive. Already the selection of projects that we have presented illustrates the interest of the theranostic approach, and we can recognize once again the productivity of the nanosciences and what they can contribute to this field. All this makes the area of theranostics very seductive for analysts and investors, but for the moment most of these projects remain in the development stage. Once again, we can close the chapter by asking whether nanomedicine as represented in the burgeoning field of theranostics is destined to remain a domain of promises for the future, or whether it will substantially transform medicine in the clinic over the years to come.

3.11 TO GO FURTHER

For more on theranostic concepts, see Hermann et al. (2017), while a general introduction to personalized or precision medicine can be found in Krzyszczyk et al. (2018) and Ryu et al. (2014). The limits of theranostic precision are explored by Yaari et al. (2016) and the applications in cardiology by Zaiou and El Amri (2017). Ligand-targeting in theranostics is discussed in Yao et al. (2016) while Meyers (2020) offers an overview of translational nanomedicine.

Herrmann, K., Larson, S. M., & Weber, W. A. (2017). Theranostic concepts: More than just a fashion trend-introduction and overview. *Journal of Nuclear Medicine: Official Publication, Society of Nuclear Medicine, 58*(Suppl 2), 1S–2S. https://doi.org/10.2967/jnumed.117.199570

Krzyszczyk, P., Acevedo, A., Davidoff, E. J., Timmins, L. M., Marrero-Berrios, I., Patel, M., White, C., Lowe, C., Sherba, J. J., Hartmanshenn, C., O'Neill, K. M., Balter, M. L., Fritz, Z. R., Androulakis, I. P., Schloss, R. S., & Yarmush, M. L. (2018). The growing role of precision and personalized medicine for cancer treatment. *Technology, 6*(3–4), 79–100. https://doi.org/10.1142/S2339547818300020

Meyers, R. A. (Ed.). (2020). *Translational nanomedicine*. John Wiley & Sons.

Ryu, J. H., Lee, S., Son, S., Kim, S. H., Leary, J. F., Choi, K., & Kwon, I. C. (2014). Theranostic nanoparticles for future personalized medicine. *Journal of Controlled Release: Official Journal of the Controlled Release Society, 190*, 477–484. https://doi.org/10.1016/j.jconrel.2014.04.027

Yaari, Z., da Silva, D., Zinger, A., Goldman, E., Kajal, A., Tshuva, R., Barak, E., Dahan, N., Hershkovitz, D., Goldfeder, M., Roitman, J. S., & Schroeder, A. (2016). Theranostic barcoded nanoparticles for personalized cancer medicine. *Nature Communications, 7*, 13325. https://doi.org/10.1038/ncomms13325

Yao, V. J., D'Angelo, S., Butler, K. S., Theron, C., Smith, T. L., Marchiò, S., Gelovani, J. G., Sidman, R. L., Dobroff, A. S., Brinker, C. J., Bradbury, A. R. M., Arap, W., & Pasqualini, R. (2016). Ligand-targeted theranostic nanomedicines against cancer. *Journal of Controlled Release: Official Journal of the Controlled Release Society, 240*, 267–286. https://doi.org/10.1016/j.jconrel.2016.01.002

Zaiou, M., & El Amri, H. (2017). Cardiovascular pharmacogenetics: A promise for genomically guided therapy and personalized medicine. *Clinical Genetics, 91*(3), 355–370. https://doi.org/10.1111/cge.12881

3.12 REFERENCES

Barbosa, Á. R. G., Amaral, B. S., Lourenço, D. B., Bianco, B., Gushiken, F. A., Apezzato, M., Silva, J. F., Cunha, M. L. da, Filippi, R. Z., Baroni, R. H., Lemos, G. C., & Carneiro, A. (2022). Accuracy of 68Ga-PSMA PET-CT and PET-MRI in lymph node staging for localized prostate cancer. *Einstein* (Sao Paulo, Brazil), *20*, eAO6599. https://doi.org/10.31744/einstein_journal/2022AO659 9

Boorse, C. (1975). On the distinction between disease and illness. *Philosophy and Public Affairs, 5*(1), 49–68.

Harbeck, N., Burstein, H. J., Hurvitz, S. A., Johnston, S., & Vidal, G. A. (2022). A look at current and potential treatment approaches for hormone receptor-positive, HER2-negative early breast cancer. *Cancer, 128*(Suppl 11), 2209–2223. https://doi.org/10.1002/cncr.34161

Mousa, S. A., Bawa, R., & Audette, G. F. (Eds.). (2020). *The road from nanomedicine to precision medicine*. Jenny Stanford Publishing.

Napoli, G. C., Chau, C. H., & Figg, W. D. (2022). Single whole genome sequencing analysis blazes the trail for precision medicine. *Cancer Biology & Therapy, 23*(1), 134–135. https://doi.org/10.1080/15384047.2022.2033058

Oudshoorn, N. (1994). *Beyond the natural body: An archaeology of sex hormones* (1st ed). Routledge.

Pucci, C., Degl'Innocenti, A., Belenli Gümüş, M., & Ciofani, G. (2022). Superparamagnetic iron oxide nanoparticles for magnetic hyperthermia: Recent advancements, molecular effects, and future directions in the omics era. *Biomaterials Science, 10*(9), 2103–2121. https://doi.org/10.1039/d1bm01963e

Roytman, M., Kim, S., Glynn, S., Thomas, C., Lin, E., Feltus, W., Magge, R. S., Liechty, B., Schwartz, T. H., Ramakrishna, R., Karakatsanis, N. A., Pannullo, S. C., Osborne, J. R., Knisely, J. P. S., & Ivanidze, J. (2021). PET/MR Imaging of somatostatin receptor expression and tumor vascularity in meningioma: Implications for pathophysiology and tumor outcomes. *Frontiers in Oncology, 11*, 820287. https://doi.org/10.3389/fonc.2021.82028 7

Simon, J. (2021). Disease diagnosis and treatment; could theranostics change everything?. *Medicine, Health Care, and Philosophy, 24*(3), 401–408. doi:10.1007/s11019-021-10015-6

Strianese, O., Rizzo, F., Ciccarelli, M., Galasso, G., D'Agostino, Y., Salvati, A., Del Giudice, C., Tesorio, P., & Rusciano, M. R. (2020). Precision and personalized medicine: How genomic approach improves the management of cardiovascular and neurodegenerative disease. *Genes, 11*(7), E747. https://doi.org/10.3390/genes11070747

Theodosiou, M., Sakellis, E., Boukos, N., Kusigerski, V., Kalska-Szostko, B., & Efthimiadou, E. (2022). Iron oxide nanoflowers encapsulated in thermosensitive fluorescent liposomes for hyperthermia treatment of lung adenocarcinoma. *Scientific Reports, 12*(1), 8697. https://doi.org/10.1038/s41598-022-12687-3

Wang, B., Tang, M., Yuan, Z., Li, Z., Hu, B., Bai, X., Chu, J., Xu, X., & Zhang, X.-Q. (2022). Targeted delivery of a STING agonist to brain tumors using bioengineered protein nanoparticles for enhanced immunotherapy. *Bioactive Materials, 16*, 232–248. https://doi.org/10.1016/j.bioactmat.2022.02.026

Wiesing, U. (2019). Theranostics: Is it really a revolution? Evaluating a new term in medicine. *Medicine, Health Care, and Philosophy, 22*(4), 593–597. https://doi.org/10.1007/s11019-019-09898-3

4 Health Under Surveillance

While exploring the area of theranostics in the previous chapter, we saw how ligands could be used to detect receptors associated with specific cells (particularly cancer cells) and so guide other functional elements to these cells for the purposes of imaging and treatment. We raised the question of whether this could still be considered diagnosis in the traditional sense of the term, and underlined the fact that this area of nanotheranostics* fitted into the logic laid out by the precision medicine movement, which has recently become a major orientation in medical research. In this chapter, we will be considering other contributions of nanosciences to diagnosis (and, secondarily, to treatment) that have the potential to change our conception of health and disease, as well as our image of a normal or healthy human body. On the side of diagnosis, we will start by looking at the enormous potential of microarrays in the investigation of the genetic contribution to both disease and a patient's response to treatment. Unlike much of the nanotechnology we discussed in the last chapter, microarrays are not a speculative projection into the future, but a technology that is solidly in place and one that only looks like it will grow in importance, notably expanding its reach from genetics to proteomics* and even to metabolomics*. The only limits to the expansion of the principles of this kind of analysis are the ingenuity of scientists and engineers in adapting and deploying the techniques as well as the capacity of computers to process and analyze the data that they generate. Furthermore, as with theranostics, analysis using microarrays is one of the driving forces behind the development of precision medicine.

After presenting the principles behind microarrays, we will discuss the paradoxical inversion of scale brought about by these analytical techniques and how the continued miniaturization of this genomic—and now increasingly transcriptomic* and proteomic*—approach to investigating health and disease risks is vastly expanding the digital data on health and disease that will be available for analysis. The reader will already recognize this as one of the major themes of this book; the production of large amounts of personal health data, particularly genomic and proteomic data, the control over this data, and more generally the relationship between an individual and the health data her or his body generates and transmits in the context of these nanomedical systems. Indeed, this leads us back to a more speculative vein where we will consider the potential for the approach underlying microarrays to be sufficiently miniaturized to be directly introduced into the human body in the form of nanoparticles. What could result from the transformation of the techniques of microarrays into the direct use of these kinds of genetic or proteomic probes (Wang et al., 2022), particularly in terms of control over the information produced by such techniques? Are we ready, conceptually, for our metabolism to be monitored in real time from within our

DOI: 10.1201/9781003367833-4

own bodies? The uncontrolled communication of this kind of personal information is the subject of many a dystopian nightmare, but this orientation of nanotechnology is going to oblige us to confront this issue of privacy as well as all the questions that circulate around the use and control of this data (Bourke & Bourke, 2020). It is, it seems, only a matter of time before people have nanoparticles inside them communicating information to all those on the outside equipped to receive it. But let us start with a much less speculative technology, one that has very much established itself in the modern molecular biology research laboratory: the microarray.

4.1 IN VIVO, IN VITRO, IN SILICO: THE INVERSION OF SCALES IN GENETIC ANALYSIS

The microarray* is today a tried and tested analytical tool providing a rapid and accurate means for determining the presence and expression of different genes or identifying specific gene mutations in a tissue sample. The technique is often used in a comparative mode, measuring differential expression between healthy tissue and tumors, for example. Laboratories all over the world are now equipped to make these analyses, using standardized machines that are increasingly purchased from international manufacturers as part of analytical packages including Agilent, Nanosys, Ayoxxa, SonoPlot, Applied Microarrays, FlexGen and Affymetrix. Thus, the microarray has already graduated from being an experimental laboratory technique to the status of a mainstream analytical tool distributed across the globe. This move from the experimental and precarious status of a new technology to a well-accepted everyday use has earned its own term in science studies: 'blackboxing'. Simply put, when a technique is blackboxed, the scientists and technicians who use it no longer need to master the principles of its functioning, but rather, in this case, can place the sample in the tray and read the results off at the end (Latour, 1987). The theory, although familiar to most users of the technology in its broad outlines, falls into the background as its use in the laboratory becomes routine.

The biochemical property behind this technique is the specificity of the bonding of bases in DNA. Because adenine binds preferentially to thymine (A-T) and guanine to cytosine (G-C) microarrays can identify specific sequences of bases using a 'probe*', which is a complementary sequence of bases located on the grid of the microarray. While these probes can be prepared by isolating the sequence from single-stranded DNA derived from the target material, they are increasingly being constructed artificially, using nanotechnology to build up 'synthetic' oligonucleotides* base by base. This technique delivers a precise pre-planned sequence of bases that can then be varied to produce a group of related base sequences. Whether they are PCR-amplified or in situ-synthesized by photolithography, these sequences of bases constitute the probes, which make up the functional part of the microarray. Thus, although the microarray analysis* does not yet operate itself at the nano level (the slides with their grids are only a few centimeters in length, but the machine to read them occupies a laboratory benchtop), it already has a claim to be included under nanomedicine* thanks to the nanotechnology integrated into its conception.

The term microarray refers to the material base for the genetic analysis, as the different probes are laid out on the surface of a slide[1] in a precise grid, and it is this physical arrangement that allows the identification of any match that is made between a probe (the oligonucleotide discussed earlier) and its target (the complementary sequence of bases) in the sample under test (Mohr et al., 2002). Thus, it is the distribution in space within the array that allows the identification of each individual probe. The key step is the highly specific combination of the probe with its complementary sequence of bases in the target—the process of hybridization. Combined with the specificity of the hybridization due to complementary base-pairing, the distribution of the probes in the array in turn allows the confirmation of the presence of the complementary sequence of bases in the sample under study. If you think of an array as like a chessboard, and the probes as being laid out one by one within the squares that make up the board, then any probe on the grid can be located by its coordinates. This means that the researcher can identify each probe by its location, which is a 'spot' on the surface. The distribution in space becomes the analytical basis for the results of a test using a microarray, and with tens of thousands of such spots on the surface of a single slide, the results have to be read and analyzed by computer rather than by visual inspection. This is, of course, the source of the analytical power and the considerable potential of this technique, as well as the large quantities of data it generates.

The subject of the test using this kind of microarray is a sample of genetic material, extracted from a healthy cell or a tumor and manipulated to produce single-stranded DNA*. Amplified reverse transcription* allows the multiplication of the number of target aRNA* or cDNA* molecules in each sample sufficiently for the hybridization of sites on the chip, allowing researchers to be confident about the results obtained from the test. The sample of genetic material is also tagged with a fluorescent dye*, and, when comparing two samples, two different colors of label can be used, one for each sample. Exciting the fluorescent labels with lasers of different wavelengths generates differentiated electromagnetic signals that register the areas of hybridization between sequences of bases. The different colors distinguish between the results from the two samples. It is the location of the probes on the slide, which, as we have already mentioned, is the key to the analysis. Following hybridization*, the sample is washed off the chip, which is then inserted into a reader that provides a combination of both fluorescent signals by interrogating each dot with laser beams detecting red and green signals. This process gives a value of the relative amount of each messenger RNA (mRNA) species (each dot) of each sample that compete for the same complementary oligonucleotide in a given dot. The laser beam checks all the dots (up to 60,000 for just a few cm^2) allowing the relative quantification of labeled aRNA* or cDNA*. The subset of expressed genes of a genome in a given situation, whether pathological or not, is called a transcriptome*.

Statistical algorithms are then used to process this data, notably treating the different colors associated with the different samples in a comparative test, allowing

[1] Various media are used for microarrays. We use the term slides, as though these arrays were all placed on glass slides, but the arrays can also be deposited on silicon chips or polystyrene beads.

scientists to see which genes are up- or down-regulated (i.e., present in higher or lower concentrations, respectively) in the samples that are being compared. This use of the technique of micro-assay analysis to compare gene expression is one of the most important today, particularly in connection with precision medicine, which, as we explained earlier, seeks to identify significant differences in order to exploit them in the context of differential, gene-specific responses to treatment. It is worth noting that similar data about gene expression can be obtained by large-scale sequencing or by High Throughput Sequencing (HTNS)* of expressed mRNA following its reversion to cDNA. This method is now widely used and its cost has drastically decreased over time.

One of the challenges for this analytical technique of micro-assay analysis as it becomes ever more widespread is the standardization of the processing of results so that they can be exchanged between different laboratories. Nevertheless, with the MIAME (Minimum Information About a Microarray Experiment) and MINSEQE (Minimum Information About a Next-generation Sequencing Experiment) systems in place, it is hoped that over time laboratories will be able to compare results more reliably.

Microarray analysis* raises many issues, not only on its material and practical side but also viewed from an ethical perspective. The use of this approach would become all the more challenging in terms of practical applied ethics if the probes could be detached from the microarray, and the techniques for identifying specific mutations or proteins could be used independently of the laboratory material described earlier. The use of single oligonucleotides as probes integrated into a molecule with the capacity to signal the hybridization with its complementary sequence of base pairs would mean that the generation of personal data concerning mutations and proteins would feed into a whole range of ongoing debates about the rights of an individual with respect to her or his biological material and the information that can be obtained from it. Another major issue today is that of security, as these techniques confront healthcare institutions and other actors with a pressing need to protect the sensitive data about a patient's health from being pirated or, more generally, from being misused.

Furthermore, with the continued miniaturization of these genetic tests, the amount of data being generated increases proportionately, with each generation of microarrays capable of testing for more sequences and generating more data. The multiplication of these tests might also lead to their losing some of their sense, but this is a problem more generally associated with the automatization and multiplication of tests. Correlation serves as the basis for establishing causal relationships, but the availability of inexpensive automated tests means that correlations will become more and more common, while the causal arguments that lie behind the conclusions that one could draw become less and less clear. Certain correlations have been discovered by accident and no one believes that they are the result of a causal connection. For example, a 2008 parodic study published in the *British Medical Journal* argued that the pope is more likely to die in a year when the Welsh rugby team wins the Grand Slam, but there are good reasons to think that the Welsh victories and the death of certain popes should be considered a coincidence and not a causal relationship (Payne et al., 2008). The comparison of multiple variables is bound at some point to turn up a statistically significant correlation, but while an automated analysis might propose a relationship of cause and effect, that does not mean that this claim is justified.

Despite fears related to the misuse of genetic information, its appropriate use could save many lives, provided that the relevant information is made available to teams working on the techniques of precision medicine. The Gene Expression Omnibus (GEO)* is the largest transcriptome depository available to date, and it is a leading example of how to use this data constructively, principally to establish disease stratification*. The analysis or meta-analysis of the transcription data from a large number of patients allows researchers to rank the different diagnoses, prognoses and therapies according to the expressed biomarkers that are retrieved in such studies, for example, for inflammatory bowel disease, Parkinson's disease or Alzheimer disease. Once again, this requires handling massive amounts of data; if we consider the approximately 20,500 expressed genes of the human genome and a quadruplicate of 100 human specimens of mRNA, the resulting excel file (400 columns each with 65,000 lines) would be around 350 Mo! With such large amounts of data, effective statistical analysis that allows researchers to retrieve robust biomarkers for pathological conditions or to establish the efficacy of a particular therapy requires well-conceived algorithms run on powerful computers.

4.1.1 Probes—A Future Outside the Array?

In the microarray, the fluorescent markers signal to the researcher (via the desktop detection machine) the pattern of gene expression in the sample or samples under test. Combined with High Throughput Sequencing of the patients' genomes, the microarray analysis allows researchers to identify important mutations of target genes that can, for example, be predictive of degenerative diseases, cancers or cardiovascular disease. If nanotechnology could provide a way of attaching a signaling element to hybridized RNA or single-strand DNA sections or, a more recent technology, probes that combine specifically with certain proteins of interest, then it would be possible to dispense with the microarray machines which will, as we have already suggested, never themselves be reduced to the nanoscale. As we explained earlier, the spatial array is the principle that allows the researcher to identify which particular probes have hybridized in the microarray approach, and so indicate what genetic sequence or protein is present in the sample under test. But if the individual probe were equipped with a signaling device that could communicate information from within the human body and whose signal was triggered or altered by the hybridization of the probe's oligonucleotide with the target base sequence (in this case, it would have to be a segment of single-stranded DNA or RNA), or the hybridization between a protein and a specific enzyme-based probe, then the probe could be used independently of the microarray to detect a genetic mutation of medical interest, thereby qualifying unambiguously as nanomedicine.

Although this kind of independent genomic or proteomic probe is not yet a reality, there are already several candidates for the nanoparticles that could supply the necessary signaling technology at the nanoscale, as this branch of research in nanoelectronics has received a great deal of attention and is the target of considerable investment. Thus, there are at least two nanoelectronics components that could be used to communicate information from a hybridized probe to an external sensor: quantum dots* and photonic crystals* (PCs). Quantum dots are a nano-sized resource that behave like semi-conductors and so can be used to assemble nanoscale electronic units, which can stock and transmit information. While cadmium-based dots represent a potential health hazard, carbon-based ones are generally considered less toxic and can be developed for use in human subjects. So, can these kinds of carbon-based quantum dots be mobilized to allow the probes already in use in micro-arrays to be disengaged from the present desktop setup to operate as independent detection devices within the human body? Assuming that the threat of toxicity can either be dispelled or overcome, this combination between quantum dot-based communication devices and probes could allow genetic tests to be carried out within the patient (or even a perfectly healthy subject) and the results to be communicated directly to detection devices on the outside of the body by the appropriate device. Indeed, quantum-dot technology is already being tried out with probes of this sort ('molecular beacons' based on molybdenum disulfide) in tests for signs of cancer in blood serum (Wang et al., 2022).

Another element that could form the basis of a nano-sized emitter is the PC, a nanoelectronics element that could again be combined with the probes that we looked at above in order to identify mutations (in particular, single nucleotide individual polymorphisms—SNIP*) of medical interest. As its name suggests, a PC acts on light rather than on electrons, meaning that deploying this element for communicating information about the hybridization of a probe requires a light interaction to transmit the signal outside the body, making the project more complicated, but not impossible.

But how realistic is this kind of technology? We have seen how the probes work in the context of the microarray systems, but what about the potential of independent probes to identify SNIPs within the human body and signal their presence to an outside receptor? The approach is not as fantastic as one might imagine.

The Massachusetts Institute of Technology (MIT), for example, is developing an interesting application of nanotechnology that aims to help health workers trace vaccine coverage more accurately and reliably in parts of the world suffering from a weak public health infrastructure. Supported by the Bill and Melinda Gates Foundation, the researchers want to accompany the injection of a vaccine with the superficial placement of nanoparticles incorporating quantum dots that would form a characteristic shape. This shape that would remain at the site of the injection can be detected by a smartphone brought near to the patient's body, allowing someone to determine whether the subject has been vaccinated or not. This project proposes the use of quantum dot technology to construct the communicating nanoparticles that remain near the surface of the subject's body.

While the form of communication and identification in the MIT vaccine project is limited to a superficial nano-structure whose form can be detected directly, we can see that it is an important step on the way to functional signaling nanoparticles. This being said, the MIT project for labeling vaccinations using nanoparticles immediately aroused concern from observers. All the people injected with this vaccine will be traceable by anyone with the appropriate application on their smartphone. While there is no doubt that in the case of this particular vaccination project researchers will have as their priority the safety of any child (or adult) receiving such an injection, this kind of technology poses obvious problems in terms of the confidentiality of patient information. The ability to detect whether or not a child has been vaccinated is clearly helpful for public health workers caring for populations that are not accustomed to the administrative paperwork so characteristic of modern Western healthcare. But this obviously raises the question of privacy and wider human rights issues, notably because, once in place, the same surveillance technology (which is independent of the vaccine itself) can be easily transferred to practices that do not necessarily have the population's health as their primary goal. Many countries already restrict access to education and collective activities to children who can attest to having received certain childhood vaccines, thanks to a stamp in a health booklet. With new systems of nano-tracing, this kind of restriction could be implemented more reliably; it would put an end to forged attestations or unvaccinated people borrowing the certificate off their friends. These are some of the ways that people got around the requirement for vaccination against COVID-19 for certain pass times such as the cinema or theater. Thus, the use of nano-tracing could be very useful for effectively imposing certain vaccinations. The rapid accurate communication of this kind of information would be practical for a number of actors in the healthcare system, but is it the kind of information that people should be communicating to others without giving their explicit consent? This public health project looks like it might be the first step in a much larger and more threatening process that integrates nanotechnology into a range of techniques that police the human body. The results of laboratory tests are already the subject of numerous precautionary safeguards intended to stop them falling into the wrong hands, but it is harder to imagine what measures could be put into place to prevent people getting access to signals placed on or in the body in the form of 'communicating' nanoparticles.

4.1.2 NBIC Convergence and Nanoelectronics

In the early 2000s, the NBIC convergence was a prominent concept for researchers and investors interested in the nanosciences, and the idea resonated with a much wider public. NBIC is an acronym covering four different areas of scientific research that, it was argued, would come together to forge a determinant technology of the future: nanotechnology, biotechnology, information technology and cognitive science. The collaboration between and ultimate integration of these four domains was going to give rise to a super science or technology that certain researchers predicted would transform the world. The promoters of this convergence were, it seems, often supporters of a transhumanist philosophy, identifying this particular development as the turning point that would allow humans to transcend their vulnerable, mortal condition. Be that as it may, over 20 years later we are not yet close to this overall unification of the different fields. Nevertheless, there are increasing numbers of collaborations across the domains highlighted by the NBIC* convergence, most often pairwise between two of them. More broadly, we can observe that almost every scientific field has enjoyed some kind of input from the different areas of the nano. We will return to this theme of how the nano are distributed across disciplinary space when we consider the future development of nanomedicine from the perspective of medical specialties in the chapter that deals with regenerative medicine.

The interaction, or more precisely collaboration, that concerns us here is the one between nanotechnology and information technology. The probes (whether conceived to identify genes or proteins) comprise one nano component and are already largely available, although they would need to be tailored to the environment of the living subject, performing without the benefit of PCR. The communication contribution will be provided by nanoelectronics. As so often in nanotechnology, this collaboration would take place in a situation of disciplinary crossover in which the field of nanoelectronics would feed into nanomedicine, while research in nanoelectronics has and will continue to have its own dynamics. Nanoscale communication devices have a number of different applications, often implicated in the 'Internet of things' that we will discuss further later, all of which contribute to orienting current research. Thus, the final form of any communicating probe will depend as much on developments in other areas of nanoelectronics as on those in microarray materials and techniques. This also means that any product that integrates these nanoelectronics communication systems, whether for the ends of health or security, for example, will raise issues touching on the production and sharing of potentially sensitive information that are common to all the other domains where nanoelectronic sensors and signaling are currently being developed. Although liable to carry particularly sensitive information, any such health sensor will be just one in a range of such devices. The proliferation of these kinds of nano-objects will be accompanied by an unprecedented growth in the amount of information available about people and their changing environments and behaviors. So, the global question is: how can we best protect ourselves from the risks associated with this profusion and availability of personal information?

Already with micro-arrays, we had noted how the constant diminution in size of the sites for the probes multiplies the number of probes in an array of the same area, producing more and more information per test. Furthermore, branching out from base sequences to proteins and SNIPs only adds to the data generated by these systems. The relationship between miniaturization and the augmentation of the amount of information suggests that reduced to the nano level the range of probes issuing from micro-arrays would be able to furnish even more digital data. Each probe seeking out a particular gene or protein will be able to communicate the presence, the absence or the relative level of their target. But, of course, the issue is not just the increase in the amount of data; in these cases, we are confronted with potentially sensitive information about the person carrying the probe: a predisposition to a cancer, a particular genetic aptitude for some task or a hereditary disease, to give just some examples. Thus, democratic societies will be faced with an increasing range of elements where the community has to decide what is acceptable in light of what becomes feasible.

The nanosurveillance of a person's health is in many ways simply an extension of health surveillance techniques that are already in place in contemporary societies, at least in those rich enough to finance them. Some, like apps on smart phones or smartwatches that monitor heart rate and blood pressure, have been adopted quite willingly by a large number of clients, despite explicit warnings that they are ceding this information to private companies. Other sources of personal health data, like labs on the skin to monitor heart function, blood pressure and blood oxygenation among other variables, are more often prescribed in the context of medical treatment or diagnosis and so operate under stricter regimes of confidentiality put in place by hospitals or other institutions. The nanosciences will inevitably multiply the information that is communicated, with the 'Internet of things' promising a future where no human need intervene for a treatment to be initiated in response to an alert concerning a treatable condition. It is hard to imagine what informed consent would look like in this context. Is it conceivable, or more importantly appropriate, that someone give 'blanket' consent for any such intervention, without being aware of whether it takes place or not? And here we are still in a scenario that involves the patient's direct personal benefit. What about the simple exchange of information (blood pressure, presence of a certain mutation, etc.) that is later used commercially? What rights will subjects have over their personal data, and will they have the possibility of sharing in the profits made from research on their genome? This question of the possibility of private companies being able to make money out of someone's genome has already been raised around cell lines, but when this same question is posed around information, and even more so when it involves aggregated data, it is going to be very difficult to manage and enforce copyright. Information on human health will also have to find its place within the Internet of things. According to researchers at the ETH (*Eidgenössische Technische Hochschule*) Zürich, there will be 150 billion objects connected together in 2025 (Helbing & Pournaras, 2015), many of them communicating personal information about their owners. We can only observe that there is a pressing need for systems to control the exchange of information between these objects, enabling an effective method for protecting personal data.

The weight of predictive algorithms will also feed into this information revolution. Already with predictive programs for providers like YouTube or Spotify, it is easy to have the feeling that a computer is deciding what we like for us on the basis of aggregated data. Furthermore, although the suggestions seem to be being made in our interest, it is difficult not to have the suspicion that these choices are ultimately driven by the need for the supplier to turn a profit rather than by any authentic desire to advance the interests of the consumer.

While the vector technology discussed in previous chapters is the most advanced in terms of bringing nanomedical products to market, it might well be that health monitoring tied into genetic profiling and the use of nanoparticles for signaling health problems will prove to be more revolutionary.

The COVID-19 epidemic has perhaps given us a glimpse of the future with respect to the tracing of treatments. Consider the QR[2] codes associated with vaccination certificates put in place around the world. In France, for example, you could not go to the theater, a cinema or a restaurant without presenting the code to someone with a smart phone whose App could read the information about your vaccination status. These codes carried other potentially more sensitive information, but without uploading the reading App oneself, it is impossible to know exactly what information

[2] Quick response codes, which have largely replaced barcodes as the source of information about products.

was being read off by the waiter at a restaurant or the ticket vendor at a cinema. This system provides a model for how information can be stored and communicated first by means of a paper or screen support (with the QR code) but later by a direct signal of the type we have been discussing earlier.

Thus, the QR code that became central to the management of the COVID-19 epidemic in many countries around the world provides a foretaste for the potential applications of nanomarkers. As we saw with the MIT project, the nanotechnology is already available to signal an important element of information concerning someone's health status: whether they have received a vaccination or not. The signaling of other health information via nanomarkers, and even nanocommunication devices, offers the possibility of more reliably tracking any number of health conditions or predispositions. Once the technology is in place (like QR codes today), the numerous applications of such nanomonitoring will offer an accessible and reliable means for tracking people's health. As we have suggested, coupled with nanodiagnostic methods, this approach can even signal health problems or predispositions to disease—information that the patient may well not be aware of—meaning that computer systems will be the first to know about someone's health problem. People who feel uncomfortable with the intrusive operation of QR codes on a vaccination certificate or the MIT signaling nanoparticles technology are not going to feel any better about the possibilities of signaling probes or other nano-devices that can communicate personal health information to a third party.

In this chapter, we have moved from the well-established materials and methods of the micro-array analysis to a quite speculative reflection about how individual probes might be used to transform this technique into an authentically nano version of the surveillance of our health. While, for the moment, the autonomous signaling probe remains a speculative nano-device of the future, the ever greater production of health information, including the negative and positive results generated by a micro-array analysis, deserves our attention. The nanoscientific world will be one of the massive productions of data, doubtless based on devices and particles communicating with each other. The choices about how to regulate this world will depend as much upon the political orientation of national governments as upon the possibilities provided by nano devices.

4.2 TO GO FURTHER

For more on the use of quantum dots to record vaccinations, see McHugh et al. (2019). An example of the use of microarray analysis in breast cancer is provided by Nuyten and van de Vijver (2008) although whole genome sequencing has now come to support this approach (Napoli et al., 2022).

McHugh, K. J., Jing, L., Severt, S. Y., Cruz, M., Sarmadi, M., Jayawardena, H. S. N., Perkinson, C. F., Larusson, F., Rose, S., Tomasic, S., Graf, T., Tzeng, S. Y., Sugarman, J. L., Vlasic, D., Peters, M., Peterson, N., Wood, L., Tang, W., Yeom, J., . . . Jaklenec, A. (2019). Biocompatible near-infrared quantum dots delivered to the skin by microneedle patches record vaccination. *Science Translational Medicine, 11*(523), eaay7162. https://doi.org/10.1126/scitranslmed.aay7162

Napoli, G. C., Chau, C. H., & Figg, W. D. (2022). Single whole genome sequencing analysis blazes the trail for precision medicine. *Cancer Biology & Therapy, 23*(1), 134–135. https://doi.org/10.1080/15384047.2022.2033058

Nuyten, D. S. A., & van de Vijver, M. J. (2008). Using microarray analysis as a prognostic and predictive tool in oncology: Focus on breast cancer and normal tissue toxicity. *Seminars in Radiation Oncology, 18*(2), 105–114. https://doi.org/10.1016/j.semradonc.2007.10.007

4.3 REFERENCES

Bourke, A., & Bourke, A. (2020). Who owns patient data? The answer is not that simple. *BMJ, Comment on the BMJ Opinion Website*. https://blogs.bmj.com/bmj/2020/08/06/who-owns-patient-data-the-answer-is-not-that-simple/

Helbing, D., & Pournaras, E. (2015). Society: Build digital democracy. *Nature, 527*(7576), 33–34. https://doi.org/10.1038/527033 a

Latour, B. (1987). *Science in Action*. Harvard University Press. https://www.hup.harvard.edu/catalog.php?isbn=9780674792913

Mohr, S., Leikauf, G. D., Keith, G., & Rihn, B. H. (2002). Microarrays as cancer keys: An array of possibilities. *Journal of Clinical Oncology, 20*(14), 3165–3175.

Payne, G. C., Payne, R. E., & Farewell, D. M. (2008). Rugby (the religion of Wales) and its influence on the Catholic Church: Should Pope Benedict XVI be worried? *BMJ, 337*, a2768. https://doi.org/10.1136/bmj.a2768

Wang, J. J., Liu, Y., Ding, Z., Zhang, L., Han, C., Yan, C., Amador, E., Yuan, L., Wu, Y., Song, C., Liu, Y., & Chen, W. (2022). The exploration of quantum dot-molecular beacon based MoS2 fluorescence probing for myeloma-related MiRNAs detection. *Bioactive Materials, 17*, 360–368. https://doi.org/10.1016/j.bioactmat.2021.12.036

5 Genetic Nanomedicine

The use of nanoparticles to deliver nucleic acids (both RNA and DNA) to the human body is an area of research that merits our special attention. These techniques in genetic medicine are being developed for one of three purposes: (i) gene targeting or silencing, (ii) gene transfer and (iii) vaccination. While the use of nanoparticles for both gene targeting and transfer remains relatively unknown to the public, the recent COVID-19 pandemic has brought the use of RNA into the limelight. Everyone now knows that the groundbreaking Pfizer-BioNTech™ vaccine delivers strands of messenger RNA (mRNA*) in order to stimulate immunity against the virus. These vaccines are derived from a viral sequence of RNA found in the severe acute respiratory syndrome coronavirus (SARS-CoV-2) that codes for a part of the 'spike' protein, which has been chosen as a target because of its essential role in the functioning of the virus. The spike protein* allows the adhesion of the virus to the cell, followed by the cell's uptake of the virus, which enables the virus to make use of the cell's system of reproduction to generate large numbers of the virus. The logic behind vaccination against this disease is that the virus, whatever mutations it might undergo, could not function without this protein, and so a vaccine that prepares the immune system to detect and eliminate any elements bearing this spike protein should act as an effective barrier to the disease. Before looking at this novel medical use of RNA and other nanoparticles, we want first to consider the older techniques that use nanoparticles to deliver genes or fragments of RNA.

5.1 USING NANOPARTICLES TO INTEGRATE ELEMENTS OF NUCLEIC ACIDS INTO CELLS

The four pillars of cancer treatment are currently surgery, chemotherapy, radiotherapy and immunotherapy (not necessarily in that order, of course). Gene therapy is a new approach in oncology that aims at disrupting the genetic code of cancerous cells. For example, the DNA of the TNF-α gene* has been successfully transferred into tumor cells by exposure to ultrasound using lipid nanobubbles containing octafluoropropane gas. Specific delivery of nucleic acids (RNA or DNA) may impede the expression of certain genes that are critical for uncontrolled cell growth, a shared, if not definitional property of cancerous cells. In experimental systems, this inhibition is mainly performed by small interfering RNA (siRNA*), a short double-stranded RNA segment that disrupts gene expression by breaking up the complementary expressed mRNA in cancer cells, thereby interrupting a step considered essential for cell growth. In mice c-*myc*, an important multi-functional oncogene has been silenced using nanobubbles containing siRNA*. As DNA is negatively charged, researchers have favored cationic (positively charged) lipid nanoparticles as vehicles for its transport and delivery, with these kinds of genetic nanoparticle delivery systems currently being explored as a potential treatment for skin tumors. Another area

DOI: 10.1201/9781003367833-5

of this research concerns siRNA gene silencers*, which have the capacity to 'turn off' genes that are causing disease. These siRNA have also been fixed to carbon nanotubes, a combination that allows them to enter into the cytoplasm of cancerous cells. One example of the projected use of these gene silencers is the suppression of the expression of a gastrin-releasing peptide receptor that is overexpressed in neuroblastoma, the most common cancer of nerve tissue and the adrenal gland in babies. This kind of siRNA–nanoparticle combination has proven to be effective in vitro using cell culture and in vivo using a mouse model with a xenografted tumor. A novel kind of conjugated nanoparticle consisting of a spherical nucleic acid has been shown to be active both in vitro and in vivo against glioblastoma, a very aggressive brain tumor occurring in young adults, as it is able to cross the blood-brain barrier, a formidable hurdle for pharmacologic agents (see Chapter 6). Spherical nucleic acids are citrate-capped gold nanoparticles to which several molecules of siRNA are added, and they are capable of blocking the expression of the targeted gene, thereby inhibiting glioblastoma growth in xenografted mice.

This new siRNA technique has also been used in several clinical trials involving inoperable pancreatic cancer, advanced solid tumors, advanced metastatic pancreatic cancer, and primary and secondary liver cancer (Sharma et al., 2022). The vectors are mainly stable nucleic acid/lipid nanoparticles, composed of cationic* and neutral PEGylated* lipids. The targeted genes are those that control vascular endothelial growth factor (VEGF*), which, as we will see in more detail in a later chapter on regenerative medicine, plays an important role in the formation of blood vessels in tumors, as well as certain oncogenes* like *KRAS**, which exhibit specific mutations in invasive pancreatic cancers.

These techniques for RNA pathway interference are less than 20 years old, and yet in 2020 more than 20 conjugates were undergoing clinical trials, with hundreds of them being tested in preclinical studies. Looking at these techniques gives us a quick overview of the use of gene therapies in oncology, a field that was almost non-existent at the beginning of the millennium, and is now burgeoning and spreading well beyond the pioneering field of cancer research. The technology of interfering RNA is now being developed in the fields of metabolic and infectious disease. This ongoing research is seeking (a) to deliver the gene–nanoparticle conjugates to the targeted cancer cells, (b) to ensure the safety of these conjugates and (c) to determine the fate of these products in the body with more precision, in particular their interaction with the immune system.

Here we would like to dedicate a section to the use of aptamers* in nucleic acid robots (nubots)*. In order to understand the interest of aptamers*, we need to reflect a little more on the human immune system. Immunity is a complex and remarkable physiological function shared by all vertebrates, consisting of two different elements, the innate and the acquired immune systems. What interests us in particular is the acquired or adaptive immune system, which responds to external threats that enter the organism, protecting against these intrusions in a specific and yet adaptive fashion.

An aptamer is a sequence of nucleic acid that recognizes a specific protein in the same way that an antibody recognizes an antigen—a situation frequently encountered in nature, as proteins and nucleic acids are in close contact in various organelles like the nucleus or ribosome. Aptamers that bind tightly to thrombin can be formed into a tubular network and this aptamer–thrombin tube can then be injected into the circulation system. In the presence of nuclein*, a protein that is specifically expressed in the endothelium* of tumors, the tube opens and allows the thrombin to cause coagulation in cancerous capillaries, thereby cutting off the blood supply to the cancer. This kind of combination is part of what is called a DNA-nubot, or a NanoroBOT put together using a nucleic acid. Such futuristic applications of supramolecular constructs will certainly have their place in anti-cancer therapy over the coming decades. The concept and technology of nubots were largely made possible through the work of Jean-Pierre Sauvage, Sir J. Fraser Stoddart and Bernard Feringa, who were awarded the 2016 Nobel Prize in Chemistry for developing the first functioning molecular machines.

We have limited the examples to the field of therapy in oncology both because this is an area where there is a great deal of pioneering research and to make the presentation of this research more coherent. Nevertheless, the use of nucleic acids combined with nanoparticles is being explored in every domain of medicine, including dermatology, neurology, nutrition and metabolic medicine. This use of nanoparticles has had particularly high-profile applications in the constantly renewed battle against infectious disease. The recent COVID-19 pandemic has thrown an unprecedented spotlight on the use of nanomanufactured vaccines and will certainly boost further research in this area, which will no doubt have knock-on effects in therapy in oncology as well.

Searching the US database (clinicaltrials.gov) in May 2021, we retrieved 671 clinical trials, including 50 completed studies on gene transfer for the gene therapy of a number of different diseases. The vectors used for this gene transfer were mainly viruses (54 using a lentivirus*, 85 an adenovirus*, and 272 where the virus was not specified), and only eight used a nanoparticle or polyethylenimine* to transport the nucleic acid. Standardized gene transfer is performed ex vivo by culturing patient cells and then transfecting* the gene of interest. Afterwards, the transfected* cells are injected into patients in the hope of restoring gene function. Gene transfer uses viral vectors, which have first been rendered incapable of reproducing themselves, in order to introduce the gene of interest into the patient's cells. This technique lies, strictly speaking, outside the scope of our book so we will not take this analysis any further. Even though viruses are clearly nano in terms of their size, their biological origin takes them out of our immediate domain of interest.

However, a new concept involving nanoparticles is currently being rolled out, one that could well mark a milestone in the history of medicine, namely, the vaccination against diseases (not only infectious ones) using messenger RNA (mRNA). These mRNA segments are copies of genes (sequences of bases in the DNA of the antigen) that are transferred to the host in order to produce the antigenic* proteins in situ. This approach represents another nail in the coffin of the 'central dogma' of molecular biology that predominated in the field of genetics in the 1960s and 1970s. This central dogma proclaims a unidirectional vision of the causal relationship between DNA and proteins. The idea behind this dogma, formulated immediately after the discovery of the structure of DNA, is that the production of proteins is a one-way causal chain, with DNA producing RNA, which acts as a sort of template for the manufacture of the proteins coded for in the DNA. This mechanism was considered to be the key to heredity. Sharing a certain sequence of bases in your DNA with your mother, for example, meant that a particular messenger RNA would be produced that would code for the proteins responsible for the blue eyes that you share with her. Nevertheless, as the history of genetics has shown over and over again, the reality of heredity and the functioning of our genetic material are far more complex in a number of different ways. First, the path from DNA to protein is neither direct nor unequivocal, but, more importantly for us here, the sequence of cause to effect does not just run from DNA to RNA, but can pass in the opposite direction as well. In particular, certain forms of RNA can change the DNA, inverting this key step. A recent and striking proof of the action of this 'reverse' flow of information from RNA to DNA over time is the discovery of the presence of endogenous retroviruses (ERV*) in the DNA of a wide range of organisms. Somewhere from 1% to 10% of your own DNA was deposited in the cell line of your ancestors over time by retroviruses like HIV that integrated a portion of their RNA into their host's DNA during an episode of infection. The philosophical implications of this phenomenon are important and have yet to be fully explored by specialists in the area, but this interrogation is already taking us a long way from the heart of nanomedicine and so we will leave this question of ERVs here. All we need to retain for what follows is the use of RNA to change the structure of DNA in a cell, with the possibility of 'programming' the cell of a patient to carry out instructions encoded by this artificially introduced base sequence. This is important, as it is a phenomenon that is becoming more and more prominent in genetic medicine and doubtless has an even more impressive future ahead of it over the coming decades. This approach is well exemplified by the attempt to propose a vaccine against gut epitope* tumors using specifically engineered mRNA transported to the host's cells by polyethylenimine*, polyethyleneglycol* or simply lipid NPs that are injected into the bloodstream or the muscles in order to 'reprogram' cells to produce the protein of interest. A team around Steven A Rosenberg and Moderna™ at the National Cancer Institute has recently published an excellent example of personalized cancer vaccines for patients with advanced-stage metastatic cancers (Cafri et al., 2020) (clinicaltrials.gov; NCT03480152).

5.2 THE USE OF NPS FOR VACCINATION AGAINST INFECTIOUS DISEASES

Until about 20 years ago, there was a consensus that the best way to elicit an immune response in humans was to use proteins, as with the tetanus anatoxin*, which can protect its recipient against this dangerous disease for decades. Thus, specific proteins or suitable parts of them have been used as vaccines without even the hypothetical risk of re-activation that haunts the use of killed or inactivated viruses. These proteins have been introduced into biodegradable nanoparticles in an effort to improve the delivery of the antigen, along the lines of the stealth strategy for other vectors presented earlier. Examples of this approach include the use of a protective antigen from *Bacillus anthracis** or the use of hemagglutinin*, a major antigen from the H5N1 avian influenza virus. These antigenic proteins are placed in biodegradable Polylactic Acid-Polyglycolic

Acid Copolymer (PLGA)* nanoparticles enhanced with toll-like receptor (TLR) ligands*, triggering immune response in cells in the three main human interfaces with the environment: the skin, the lungs and the intestines. These supramolecular constructs are similar to viruses in size and shape, making them nanoparticles, and, as we have already mentioned, they do not even theoretically have the potential to replicate in cells. These products of bioengineering have been shown to stimulate both antibody production and cellular defense based on T-lymphocyte* immunity. In these cases, the antibody response has been shown to be long-lasting in mice, as is the case with the vaccine against yellow fever that was developed following a more empirical route.

This 'classical' protein-based approach to designing vaccines has recently been eclipsed by the rapid progress of messenger-RNA based techniques. In order to understand the accelerated progress in this field in recent decades that this new technique represents, we need to look more closely at the development of genetic approaches to vaccines.

In 1990, Jon Wolff and his team published an article in *Science* (Wolff et al., 1990) that brought to the attention of the scientific community the proof of concept that an intramuscular injection of a DNA plasmid* can enable the induction of a complete immune response in rodents, specifically an antibody and cellular (T-cell*)-mediated response. The vaccinology community was enthusiastic about this development because DNA is the most stable form of nucleic acid and is easy to produce and to purify, giving a homogeneous product that can be used safely. The potential of this approach has been further enhanced by the demonstration of the effectiveness of an RNA vaccine in a mouse tumor model. Nevertheless, although this anti-cancer vaccine has been shown to work in small animals, the approach has not translated successfully into more complex species, in particular humans, despite a large number of studies conducted over recent decades. In the meantime, since their discovery at the end of the 1980s, lipid vectors have been improved by the use of cationic* lipids and their formulation into lipid nanoparticles (LNPs) (Ryals et al., 2020). In 2018, the FDA approved the first RNA-LNP vaccine for the treatment of inherited proalbumin amyloidosis*. As recently as 2020, no nucleic acid-based DNA vaccine for human use is mentioned in any of the three main pharmacopoeias. The emergence of the COVID-19 pandemic in 2020 has radically changed the situation; money and resources were rapidly made available to organize and execute both preclinical and clinical studies, leading to the production of literally billions of doses of the innovative RNA vaccine in 2021. Indeed, Britain, recently separated from the European drug approval mechanisms, approved the first RNA vaccine in 2020, following clinical trials that had taken less than a year to put in place.

Thus, the COVID-19 pandemic has accelerated the introduction of the mRNA vaccine, and by the end of 2021, the mRNA-based vaccines produced by Pfizer-BioNTech or Moderna are likely to be the most widely used manmade nanoparticle construct in the world, with maybe as many as two or 3 billion humans receiving an intramuscular injection of the product.

Before describing these vaccines in detail, we should note that RNA vaccines can be divided into two types: mRNA vaccines and self-amplifying RNA replicons. The former use the messenger RNA to produce the desired protein, while the latter are based on a synthetic virus that replicates only one single gene—the one that is of interest for the vaccination.

What are the advantages of RNA vaccines? Unlike DNA vaccines, they are not even theoretically capable of integrating any gene or even any smaller sequence of bases into the host genome and so do not risk altering the genetic makeup of the patient. The RNA is also easy to obtain in a purified and homogeneous form, a considerable advantage over the protein that they are intended to produce. In addition, RNA can be translated* directly into the encoded protein* by the cytoplasm of the cell, while DNA has to enter the nucleus of a cell in order to be active. There is another less foreseeable advantage—these RNA structures trigger a more vigorous immune response (especially innate response) than their DNA counterparts (Hogan & Pardi, 2022). However, these RNA vaccines suffer from a major drawback: the spontaneous hydrolysis of the RNA molecule due to its own structure. This problem is inherent in these agents, as they necessarily contain ribonucleotides*, which are susceptible to the action of enzymes called RNAses that decompose the RNA. This constitutive feature is responsible for the biggest practical obstacle faced by the first RNA vaccines against COVID-19. In order to prevent this decomposition from taking place, the Pfizer-BioNTech vaccine needed to be stored at 80°C below zero. However, over time, this problem has been attenuated by using modified ribonucleotides* that are less sensitive to the action of these enzymes.

The main vaccines against SARS-CoV-2 virus used around the world are based on the *S* (Spike) protein present in multiple copies on the corona (crown) of the virus, which gives it its name. mRNA were obtained by genetic engineering using a bacterial plasmid in which the S gene was cloned and two sequences, namely, one coding for the cap and another for the poly-A tail, were added to each end of the molecule to improve the translation in the recipient's cells. The mRNA was further purified to render them homogeneous before being encapsulated in lipid nanoparticles (LNPs), ready for intramuscular injection. To date, two RNA vaccines are used around the world: mRNA-1273 developed by Moderna™ in collaboration with the National Institute of Allergy and Infectious Diseases and BTN162 developed in parallel by BioNTech working with Pfizer.

The mRNA-1273 vaccine is composed of lipids (glycerolipids, phosphocholine and myristate derivatives) certain of which form a complex with a polyethylene glycol (PEG)* buffer in order to guarantee an optimal pH as well as with sucrose to stabilize the polymer during the freezing process. The vaccine that results from this approach is similar in form to a lipid nanoparticle of approximately 100 nm in size. The composition of the BNT162b2 vaccine is quite similar, including four lipids: a cationic* one, a PEGylated* one, a salt to adjust the ionic strength* and a buffer to fix the pH*. This combination of the cationic* and the PEG* lipids allows the vaccine to fix the anionic* charges of the nucleic acid molecules, and at the same time increases the production of the protein in the cytoplasm of the targeted cell following the uptake of the RNA by endocytosis*.

Both the mRNA vaccines used in the early 2020s for protecting patients against COVID-19 have their uridine (one of the four nucleosides that make up RNA) replaced by methylpseudouridine in their composition. This substitution makes them more resistant to enzymatic* hydrolysis* within the cells and increases the probability of their being translated into the proteins that will trigger the appropriate immune response.

The introduction of one of these vaccines into the patient's muscle (generally at the level of the shoulder) provokes a local inflammation that attracts immune cells to the site of injection. The liposomal nanoparticles conjugated with the mRNA are then able to cross the membranes of the muscle cells (myocytes*) and enter the cytoplasm where the coding of the messenger RNA can be translated into proteins by the polysomes of those cells. Once again, this approach avoids the risks associated with interaction between the vaccine and the patient's DNA within the cell nucleus. The proteins coded for by the mRNA and produced in the cytoplasm of the host's cells are fragments of the S protein of the Coronavirus that are recognized as 'non-self' by the patient's macrophages and dendritic cells*. This interaction (which casts the S-protein in the role of the antigen) induces the production of antibodies, and the vaccinated subject retains this capacity to react to the Coronavirus in the form of T-lymphocyte*-cell-acquired immunity. It is perhaps surprising that this relatively simple mechanism for immunization was only developed by biologists in the 1990s, particularly as it is much easier and quicker for a molecular biologist to produce large amounts of purified mRNA than purified protein. We can maybe remark with Paulo Coelho that 'It's the simple things in life that are the most extraordinary; only wise men are able to understand them', but this being said, everything seems easy to predict when we recount history retrospectively. Be that as it may, it seems clear that from now on, a whole range of innovative vaccines will be based on this new RNA approach.

As an illustration of how rapidly things are changing in this field, we only need to consider the replicon* technology that is already transforming the conception and functioning of these prophylactic medicines. Indeed, the next generation of RNA vaccines is likely to be dominated by these replicon- or 'self-replicating RNA' vaccines. In molecular biology, a replicon is the name given to a piece of nucleic acid (RNA or DNA) that is able to make copies of itself. A self-amplifying RNA (saRNA) can reproduce itself hundreds of times over before the stage where the RNA enters the cytoplasm, thereby increasing the capacity of the same dose of vaccine to generate the proteins that will act as antigens to trigger the desired immune response. The overall result is a kind of biological amplification in the host cells that will doubtless find medical applications beyond this field of vaccines as well.

Of course, the saRNA is a much larger and more complex molecule than the mRNA in use in today's vaccines. Nevertheless, in spite of the insertion of genes that ensure the self-replication of the nucleotide, the molecule's capacity to re-activate as an infectious virus is once again absent because the genes coding for the structural proteins of the virus are missing from the engineered genome. The architecture of these kinds of saRNA viruses remains similar to that of the mRNA vaccines used in the recent COVID-19 pandemic. The long replicon (saRNA), which constitutes the vaccine's cargo, is a negatively charged molecule and is accompanied by a selection of physiological fluids. To ensure chemical stability, the RNA segment needs to be surrounded by cationic lipids, meaning that it is first coated with chemicals like PEG*, alginate* or hyaluronic acid*, a natural component of tissues. The combination of all these components gives a spherical LNP. As with the first RNA vaccines, these replicons contain a modified molecular cap and tail to lengthen their lifespan in the host cells so that they are readily recognized as 'non-self', which results in their being internalized by the dendritic* cells that constitute the first line of human immune response. A replicon is always derived from a 'sabotaged' virus genome to avoid the production of virus progeny during the replication* stages. Here, the goal is only to produce the protein of interest that will act as the antigen (like the S protein in the case of COVID-19) and not replicate the virus, the mechanism at work in a viral infection. After delivery to immune competent cells like macrophages and dendritic cells, the translation machinery of those cells allows the production of not only the encoded protein (in the case of COVID-19, the S protein) but also the polymerase*, an enzyme specific to that replicon able to reproduce the RNA many times over. These dendritic cells will then present the S protein to the lymphocyte B and T cells, triggering the production of specific antibodies and the destruction of any cell that presents this protein (in this case, the spike protein). This is how vaccination with messenger RNA leads to T cells destroying any cells producing S proteins, thereby preventing the implantation of the Coronavirus in the body.

Clearly, the second generation of RNA vaccines is on its way, in the form of products based on saRNA molecules. The saRNA will, it is hoped, not only make the vaccine more effective but also shorten the development time for updated versions capable of blocking new strains of any given virus. This approach opens up a range of possibilities for responding to rapidly mutating viruses, not only Coronavirus but also other RNA viruses that can transfer from species to species, such as influenza viruses or human immunodeficiency viruses (HIVs). The genomes of these RNA viruses are not as stable as those in ones based on DNA, leading to more mutation and so a more rapid evolution, allowing them to by-pass widely used vaccines more readily, as well as quickly becoming resistant to therapeutic treatments like antibiotics. Although the flexibility of the RNA approach will be valuable for keeping up with dangerous variants, this kind of 'immunological arms race' raises its own problems, just like the other therapeutic and vaccinal innovations that have marked the history of medicine.

Despite the current interest in mRNA vaccines, their protein-based counterparts should not be considered redundant. The COVID-19 pandemic triggered significant progress in this area as well. Mosaic NPs—a kind of synthetic virus without any genome—were engineered to fix and present proteins of interest on the surface. For this, scientists used a part of the spike protein (drawn from the receptor binding domain) extracted from eight varieties of coronaviruses. These engineered proteins were fixed on the surface of the mosaic NPs like the spike protein of a native coronavirus. These NPs elicit immune response against the eight different coronaviruses and surprisingly even against others that were not represented in the mosaic NP. These kinds of mosaic NP should help in the development of preventive vaccines, allowing the immune system to 'learn' to recognize invariant epitopes of a protein subjected to evolutionary changes.

5.2.1 How Vaccines Fit Into Nanomedicine

The hepatitis C virus (HCV), like every other virus, is a nanoparticle. But how can we best fit viruses into our presentation of nanomedicine? Right from the outset we can note that viruses are conceptually delicate objects to handle, posing particularly thorny questions about classification in the domain of the living. While clearly part of the living world, viruses have generally not been considered to be living organisms by biologists, or even to be alive. Unlike bacteria, they are not capable of surviving and reproducing without the 'help' of their host, the living organism into which they are introduced (for a discussion, see Villarreal (2004)). Already in 1962, the French Nobel laureate André Lwoff could make the comment 'Whether or not viruses should be regarded as organisms is a matter of taste' (Lwoff, 1962). While obviously meant to be humorous, the comment reflects a deep movement in biology that has turned away from the formerly 'hot' topic of the frontier between the living and the non-living. Nevertheless, while this debate has all but disappeared from biological research, it has continued in philosophy. To reflect this intellectual shift, we will not comment any further on the question of whether a virus is alive or not, but whatever one's answer to this question, it seems that viruses are nano in scale, hence their inclusion as vectors in nanomedicine. Thus, while viruses are generally measured as being tens to hundreds of nanometers in diameter, we will now turn to the interesting case of the hepatitis C virus (HCV) the size of which raises some very interesting questions.

Among the millions of viruses present on earth, HCV possesses the unfortunate property (from a human perspective) of interfering with the healthy functioning of the body of *Homo sapiens*, in particular the liver: interference that can lead to serious health problems and even death. Thus, along with some other viruses (very few in light of the large numbers of viruses that share[1] our planet), such as the human immunodeficiency virus (HIV), the SARS-CoV-2 and various flu viruses, HCV is harmful and potentially deadly for its human hosts. Indeed, hepatitis C kills hundreds of thousands of people every year around the world, and incapacitates millions more through liver disease and its various debilitating consequences. More than 185 million people around the world are thought to be infected with HCV, although, because infection with the HCV is usually asymptomatic and there is no systematic testing program for the disease, only about 5% of these infections are likely to be identified. All this is well-known, but a less familiar feature of HCV is that it is a striking example of a non-unequivocal naturally occurring nanoparticle. It is this aspect of HCV that we want to explore and then use it to suggest a much wider issue concerning how we think about nanoparticles in the biological milieu and so a central question for the future of nanomedicine.

Unlike most other viruses, the existence and nature of the hepatitis C virus were not proven or at least initially confirmed by electron microscope imagery, but rather

[1] Here again, we see the difficulties that come with the nature of viruses; the idea of sharing seems inappropriate for a non-living object – who would say we share the planet with granite or sand? And yet, few people would contest the claim that we share the planet with viruses.

all the proof remained at the level of molecular biology. At the end of the 1980s, Michael Houghton, who was working at Chiron Corporation (Emeryville, CA), was collaborating with Qui-Lim Choo, Daniel Bradley and George Kuo in research into the form of hepatitis that was associated with neither the hepatitis A virus nor the hepatitis B virus. Logically enough, this new form of hepatitis had been dubbed hepatitis C, and the identification of the infectious viral agent behind the disease earned Michael Houghton the 2020 Nobel Prize in physiology or medicine. The researchers took six years (1982–1988) to identify the virus, and were unable to isolate and photograph their find by what by then were standard means. Instead, they had to rely on a complex experimental method of genetic deduction to arrive at the conclusion that they had successfully isolated the incriminated virus. Even after they had concluded to their complete satisfaction that they had effectively obtained the virus, they were unable to visualize it using the electron microscope. The difficulties confronting Houghton and his collaborators were due to the indeterminate nature of the virus.[2]

Even today, HCV remains very difficult to amplify in vitro, and the first electron micrographs of the virus date only from the end of the twentieth century. Indeed, it was some 27 years after its molecular discovery that the first photographs taken through an electron microscope were published. This work was performed by a French group headed by Jean-Christophe Meunier in Tours when they managed to photograph the virus following its uptake from human plasma onto microscopy grids labeled with an antibody specific to one of the proteins found in the virus' envelope. This story of the identification of HCV is not only fascinating on its own account, but it also suggests that we might need to adjust our way of thinking about the place of nanoparticles in nanomedicine. What is remarkable about the behavior of HCV under these experimental conditions is that it does not appear to possess a unique form but rather various different forms depending on the context in which it finds itself. Indeed, its form, lipid composition and dimensions vary as a function of the patient and even vary within the same patient as a function of their diet. Why or how could this be so?

It turns out that HCV integrates a variety of lipids (triacylglycerol, cholesterol) and apolipoprotein E to put together a lipo-viro-nanoparticle as part of the process of its maturation in hepatocytes. During this process, the dimensions of the virus can more than double, increasing from 60 to around 200 nm in diameter, attaining its maximum size following the ingestion of meals rich in fats, like after drinking a milk shake! This process of agglomeration gives rise to an asymmetrical virus complex without any well-defined form, but a conformation that can vary as a function of time and diet in the same patient. While this amorphous and changing form of the Hepatitis C virus posed so many problems for its identification, it is probably not as exceptional as it might appear. An understanding of the mechanisms behind this shifting form (and size) should encourage us to reflect on what lessons we might be able to learn from the case of HCV that apply to nanoparticles more generally, especially their existence in different biological media. This is a vitally important issue for nanomedicine, as all living organisms, the human body included, present a range of different organic media that may have an effect on each nanoparticle, enlarging

[2] Houghton offers a version of this story in his Nobel acceptance speech www.nobelprize.org/prizes/medicine/2020/houghton/lecture/ (accessed 14 July 2021).

or shrinking them in significant ways. Thus, the useful lesson is that it might well be an error to conceive a nanoparticle as a well-defined object of a particular size that behaves in the same way independently of its context. As soon as they leave the test tube (and even before) nanoparticles may be as much the products of their local context as their 'core' chemical composition. The chemical formula that we are accustomed to writing down, long and complicated though it may be, does not accurately portray the complex and shifting molecule that includes the acquired crown of proteins (the corona) in biological media.

This effect also leads us once again to consider the pertinence of an excessively rigorous size-based vision of what counts as nano. If a virus like HCV can vary in size between 60 and 200 nm, then with the limit of the nano fixed at 100 nm, the same virus can be a nanoparticle or not depending on such factors as the patient's diet or the local viral and bacterial environment. In order to avoid this possibility of evolutive re-classification as a function of the virus' medium, we need a definition that is a little subtler than the 100 nm rule. Coming at the issue from another perspective, this corona effect also opens up novel avenues for how to treat certain diseases, starting with viral infections. Replacing certain molecules in the corona of a deactivated virus with therapeutic agents, for example, offers a range of innovative and potentially very specific therapeutic possibilities.

So the assertion of the Nobel Prize Winner Sir Peter Medawar that 'a virus is a piece of nucleic acid surrounded by bad news' remains true: in the case of the Coronavirus responsible for the pandemic from 2019 onward, the bad news has been the spike proteins and in the case of HCV, they are unique mixture of lipids, proteins and apolipoproteins. Just as after all the evils had escaped from Pandora's box, hope remained, the bad news might well be accompanied by some good news.

The COVID-19 epidemic has caused a great deal of suffering, killing over six million people around the world. This figure would be much higher if it were not for the rapid introduction of vaccines, notably the mRNA vaccine developed by BioNTech in collaboration with Pfizer. Thus, on the positive side, the epidemic has helped to advance new RNA technologies, which, working at the nano level, have opened up a number of avenues of research for the world's vaccine producers to explore. As the different behavior of viruses and vaccines becomes better known, our knowledge of both will inform a range of techniques at the nano level. Even though scientists will doubtless continue to consider its RNA or DNA as the core[3] of a functioning virus, the example of HCV suggests that we need to take other aspects into account as well, notably its immediate environment. Increasingly, modern medicine is pushing for more holistic approaches that do not consider disease simply as the functioning of a well-defined disease entity. One of the orientations that have been promoted through precision medicine is an analysis of the exposome: the whole range of environmental factors that can influence an individual's health. Similarly, the HCV example suggests that viruses and other nanoparticles can change substantially in the passage from their in vitro isolation to their integration in a living body. In the end, the most valuable lesson these viruses can teach us is to think more about everything that surrounds nanoparticles, and which may even be thought of as constitutive, even if only temporarily.

[3] We avoid using the term 'essence' here, as it bears too many inappropriate connotations in philosophy.

5.3 REFERENCES

Cafri, G., Gartner, J. J., Zaks, T., Hopson, K., Levin, N., Paria, B. C., Parkhurst, M. R., Yossef, R., Lowery, F. J., Jafferji, M. S., Prickett, T. D., Goff, S. L., McGowan, C. T., Seitter, S., Shindorf, M. L., Parikh, A., Chatani, P. D., Robbins, P. F., & Rosenberg, S. A. (2020). MRNA vaccine-induced neoantigen-specific T cell immunity in patients with gastrointestinal cancer. *The Journal of Clinical Investigation, 130*(11), 5976–5988. https://doi.org/10.1172/JCI134915

Hogan, M. J., & Pardi, N. (2022). MRNA vaccines in the COVID-19 pandemic and beyond. *Annual Review of Medicine, 73*, 17–39. https://doi.org/10.1146/annurev-med-042420-112725

Lwoff, A. (1962). *Biological order*. MIT Press. http://archive.org/details/biologicalorder00lwof

Ryals, R. C., Patel, S., Acosta, C., McKinney, M., Pennesi, M. E., & Sahay, G. (2020). The effects of PEGylation on LNP based mRNA delivery to the eye. *PloS One, 15*(10), e0241006. https://doi.org/10.1371/journal.pone.0241006

Sharma, S., Gautam, R. K., Kanugo, A., Mishra, D. K., & Kamal, M. A. (2022). Current synopsis on siRNA therapeutics as a novel anti-cancer and antiviral strategy: Progress and challenges. *Current Pharmaceutical Biotechnology*. https://doi.org/10.2174/1389201023666220516120432

Villarreal, L. P. (2004). Are viruses alive? *Scientific American, 291*(6), 100–105. https://doi.org/10.1038/scientificamerican1204-100

Wolff, J. A., Malone, R. W., Williams, P., Chong, W., Acsadi, G., Jani, A., & Felgner, P. L. (1990). Direct gene transfer into mouse muscle in vivo. *Science (New York, N.Y.)*, *247*(4949 Pt 1), 1465–1468. https://doi.org/10.1126/science.1690918

6 Toxicology of Nanomaterials
A New Toxicology?

Toxicology is the scientific study of the adverse effects of substances (such as synthetic chemical compounds or micro-organisms and their products) and devices on living systems (macromolecules, cells, organs or the entire body), whether these potential toxins be natural or man-made.

In this chapter, we will present the basic elements of toxicology applied to nanoparticles, or nanotoxicology. As we shall see, this nanotoxicology is importantly different from other domains of toxicology. As the particles involved approach the molecular scale, and, as we have already seen, the contact surface area of materials increases dramatically, new physiological phenomena are observed that depend on the nano scale; the principal aim of this chapter is to bring these novelties to the fore. This will enable us to understand the new challenges that the nanosciences and, in particular, the introduction of new nanoparticles into our bodies and into the environment raise for toxicology, and the need to invest more significantly in this domain.

6.1 NEW METHODS, NEW POSSIBILITIES, NEW INVESTIGATIONS

As we already explained in the introduction, nanoparticles did not arrive with the nanosciences at the end of the twentieth century, as there have always been naturally occurring nanoparticles, and human activity, particularly the use of the internal combustion engine, has generated many more since the end of the nineteenth century. Nevertheless, with the development of techniques for identifying and dosing the whole range of these nanoparticles, researchers have been able to establish a specific domain of toxicology dedicated to studying and predicting the toxic effects of these particles. Of course, the development of the nanosciences has also introduced a range of new artificially produced nanoparticles into our environment that have to be taken into account in these kinds of toxicology studies. Indeed, the nanoscience is one of the rare areas in the history of technological innovation where the toxicological threat of new substances has been at the forefront of researchers' concerns. While the plastics industry initially developed in the first half of the twentieth century without a great deal of accompanying research into the toxic effects of the numerous mass-produced polymers that were put on the market, the situation has been very different for the nanosciences. Right from the time when the first new nanomaterials were being produced, toxicologists have been working to try and assess the health implications of these substances both for humans and for the wider environment. In

DOI: 10.1201/9781003367833-6

part, this is because, for the reasons we will outline in this chapter, but largely related to their size, nanoparticles are suspected of being potentially more dangerous for human health than even their microscopic forebears. But this interest in the potential dangers of the new nanosciences and their products is also the result of lessons learned from other areas of technological innovation, in particular the development of human genetic research at the close of the twentieth century.

The human genome organization (HUGO) project, a multi-billion dollar project aimed at sequencing the human genome, was launched in the late 1980s, and ended at the beginning of the twenty-first century with the first complete sequences. From early on in the government-led financing of the project, money was set aside for studying the impact of this research, channeled through a specific program (the Ethical, Legal, and Social Implications—ELSI—Program). This money financed cross-disciplinary research, featuring collaboration between researchers from a wide variety of academic backgrounds. One idea emphasized in these ELSI studies was that of anticipating problems associated with genetic engineering. The divergent attitudes of Europe and the United States to genetically modified organisms, particularly corn, suggest that these ethical studies may have been an epiphenomenon for an agricultural industry driven by economic concerns. It was not ethicists who fixed the rules, but rather governments under pressure from various lobbies. The result has been that the American market has been very open to genetically modified corn, with the conviction that the risks are negligible, while Europeans have stood against both the growth and the importation of genetically-modified corn. This reticence on behalf of the Europeans is based largely on the 'precautionary principle', which privileges the place of potential, as yet undetermined risks in the equation that assesses the balance between benefits and risks. As we have suggested these differences in approach seem to have more to do with economic factors and the political process (with lobbying an important aspect) than with the work of philosophers on the more straightforwardly ethical questions.

Despite the questionable value of the ELSI element of the Human Genome project, the US National Nanotechnology Initiative has followed its example, although not with such a high level of investment. Once again, the idea of anticipating developments (both negative and positive) has been an important part of the ethical approach to nanotechnology, and toxicology studies have figured prominently in this research.[1]

In this chapter, rather than running through a list of different nanoparticles and enumerating the suspicions about the potential health risks associated with their dissemination and use, we will be considering the physiological mechanisms that might lead to such risks of toxicity. How can these particles enter the body, and what are the potential reactions of the immune system or the tissues that make up our bodies to the introduction of these substances? What we want to bring out are the differences and specificities of nanoparticles in terms of their interaction with the body compared to other particles, and what this means for the potential toxicity of these materials.

---

1 See the National Nanotechnology Initiative, in particular the section on Responsible Development, www.nano.gov (accessed 21 July 2022).

6.2 ENTRY INTO THE BODY: THE THREE PRINCIPAL ROUTES

So, how can nanoparticles potentially affect our health? We can open this inquiry by asking how they might enter our bodies. What are the pathways for the entry of nanoparticles and nanomaterials (which we will assimilate to nanoparticles to simplify the presentation of this area) into organisms, as well as the main effects on (or dynamic interactions with) these biological systems?

Taking a global view of the interaction between organisms and nanoparticles, we need to see that these particles can enter into the body via three main routes, and once in an organism they are able to circulate within it, be transformed by specialized cells or metabolic functions, before, for the majority of them, being eliminated. This process known as 'ADME', which stands for Absorption, Distribution, Metabolism and Elimination, is standardly studied for pharmaceuticals, chemical compounds* or xenobiotic* substances in general, whether man-made or natural. After looking at how nanoparticles can be absorbed into the body, we will then go on to consider their main effects at the cellular and molecular levels. Finally, we will highlight the differences between classical toxicological and nanotoxicological studies, bringing into focus the new paradigms that need to be put in place in order to evaluate the toxic effects of nanoparticles.

To begin our overview of the toxicological potential of nanoparticles, we need to consider the three main routes of entry into the body. We will leave the intravenous route to one side, even though it might be the first to come to mind in a clinical setting, as this only concerns the deliberate introduction of nanoparticles into the body and so deserves special consideration under the head of pharmacovigilance rather than general toxicology. Leaving injection to one side, to gain access to a living human body, the nanoparticles need to cross one of its three main barriers with the environment: the skin, the lungs or the digestive system. These frontiers between the human body and its surroundings are quite extensive. If we take a 70 kg human, for example, the surface area of her or his skin is approximately two square meters, the surface area of the lungs around 70 and the intestines some 200 m^2. But before turning to these barriers between the body and the environment, it is helpful to contextualize the human body's most general defense mechanism, what might be considered its first line of defense, and that is the macrophages that circulate throughout the body. Indeed, as we saw when we examined the evolution of liposome vectors, even the latest nanoscale drug vectors described as being 'stealthy' are more or less readily identified by the body's macrophages as foreign, potentially dangerous intruders. Once the macrophages have detected them, they can then go on to absorb and eject these microscopic intruders. In skin, macrophages and dendritic[2] cells share this phagocytic function, constituting a 'phagocytic barrier' also known as the 'mononuclear phagocytic system*'.[3] Macrophages are ubiquitous cells that can be found in almost all our organs, like bone marrow, the liver, the spleen, the brain,

[2] Monocytes can turn into either dendritic cells (in tissues exposed to environment) or macrophages. Dendritic cells are presenting antigens to immune cells.

[3] This is also known as the reticuloendothelial system, or macrophage system is a part of the immune system that consists of the phagocytic cells located in reticular connective tissue. Also known as the macrophage or reticuloendothelial system, the mononuclear phagocytic system is the part of the immune system made up of the phagocytic cells found in the reticular connective tissue.

the lungs, blood, lymph nodes or in the lymphatic system. As soon as nanoparticles are internalized into macrophages and depending on their biopersistency, their components are hydrolyzed (broken down) into smaller molecules by macrophage vacuoles in a time- and chemical-dependent manner.

6.2.1 The Skin

The skin is not permeable to microspheres greater than 3 μm in diameter, but nanoparticles of 70 nm in diameter (about 50 times smaller) have been shown to penetrate the skin, notably through hair follicles and more particularly when these follicles move, as when someone flexes their wrist, for example. Thus, fluorescent nanospheres of 500 nm have been shown to penetrate to the dermis by absorption through small intercellular channels and hair follicles. The nanoparticles accumulate in the dermis much as the dyes used for tattoos do, promising to be as permanent as these tattoos, which constitute a lifelong engagement. Thus, if these nanoparticles are not biodegradable, once they have arrived at this level, they can well stay in your skin for the rest of your life.

One idea, although for the moment it has not been proven, is that certain nanoparticles can cross skin cells in the same way the Ebola virus does. This deadly disease is transmitted by a 'cylindrical' nanometric-sized virus (40 × 800 nm) that can gain access to the body's circulatory system by passing through the skin, either independently or transported by macrophages.

6.2.2 The Lungs

The penetration of nanoparticles into the body by inhalation has been widely studied for man-made nanomaterials, as it is suspected of being their major route for entry into the body, potentially leading to toxic effects. In general, there is a rule concerning all microscopic nanoparticles (such as viruses or inert particles like those found in dust or diesel exhaust), the smaller they are (in terms of aerodynamic* diameter) the deeper into the lungs they can penetrate. Brownian diffusion*, convection and sedimentation allow particles of less than 100 nm in diameter to collect in the alveoli, the deepest area of the lung where gas exchanges take place, thereby constituting a potentially serious health threat. In the alveoli, phagocytosis is assured by alveolar macrophages, which can absorb these materials. Like any other of these immune defense cells, alveolar macrophages are capable of identifying such nanoparticles as 'non-self', whether they originate from artificial or from natural terrestrial sources, such as iron and coal dust in mines, plants (although not nanometric, microscopic pollen provokes a high proportion of all allergic reactions), industrial pollution (ultrafine particles from diesel combustion, to signal a common pollutant in urban areas) or even airborne viruses and microbes like *Mycobacterium tuberculosis*, the bacterial agent that causes tuberculosis. While, as we have just explained, these nanoparticles can have a natural origin, it is the man-made particles that have particularly interested toxicologists in the last few decades. This research reflects a widespread anxiety about new materials in general, and nanomaterials in particular, but for the moment while much has been discovered about the behavior of nanoparticles in the body, no proof has been forthcoming about any definite associated health risks.

By capturing inert and living micro- or nanoparticles, alveolar macrophages can stop their spread deep into the lungs. Providing they are biodegradable, that is to say constructed out of proteins, carbohydrates, lipids or nucleic acids, these particles break down into smaller components that can be recycled by biochemical processes, thereby inactivating the offending nanoparticle. The success of the macrophage system in blocking the entry of any nanoparticle into the organism is a gauge of the protection provided against its potential toxicity. This, of course, depends in turn on the ability of these cells to recognize the nanoparticle as 'non-self' in its defense of the self that is the host organism. This is how these immune cells play their role as a barrier against potential airborne invaders, but if they are unable to fulfill this essential role, we can expect to see the rapid development of a local inflammation or infection. This overactive inflammation can in turn be the source of major complications, as is seen with the *Legionella* bacteria or with a number of viruses. Indeed, it is this kind of runaway inflammation that has made the SARS-CoV-2 (COVID-19) virus so deadly in the recent pandemic. When a microorganism is capable of replicating and producing a large number of similar infectious agents, the infection can become systemic, accompanied by fever and potentially leading to cardiovascular shock. It is easy to see how alarming these scenarios can be, and the idea that nanoparticles could trigger this kind of reaction seems theoretically plausible. But there is no evidence to suggest that any nanomaterials have yet caused such a situation, even following widespread exposure to airborne nanoparticles.

Once embarked by macrophages, nanoparticles migrate toward lymphatic nodes passing through the systemic circulation in immuno-modulating organs: bone marrow, the thymus and the spleen. An important element for removing nanoparticles from the lung is the so called 'mucociliary escalator' that allows microparticles as well as nanoparticles to be lifted up and transported by the very small 'lashes' existing at the apex of the bronchial epithelium cells and this independently of whether they have already been captured by macrophages. These lashes move their charge, thanks to a movement that has been likened to the movement of a field of wheat swaying in the wind. This physiological function can carry the nanoparticles to the pharyngeal crossroads, where they can be expectorated out of the body, or swallowed and then eliminated via the gastro-intestinal tract.

There is another troubling scenario, which is the case where the lungs become overloaded with nanoparticles, a situation frequently encountered in rodent experiments. This overload can provoke the violent inflammation of lung interstitial tissue, impeding the process of gas transfer across the alveoli–capillary barrier*, leading to chronic obstructive pulmonary disease*, generating cardiac toxicity and even inducing heart failure. Following the uptake of nanoparticles by macrophages, these nanoparticle-loaded macrophages retain their ability to migrate across barriers, allowing them to attain the lymph nodes and compromise the immune system, a situation that has already been observed with asbestos-loaded macrophages. Asbestos fibers are not nano-, but micrometric in size; they are very biopersistent* and can lead to serious lung toxicity and provoke lung cancer or mesothelioma, a specific cancer of the external lining of the lung. This kind of trans-barrier migration has also been observed with nanoparticles, some of which also display high levels of biopersistency*. Evidently, these parallels with asbestos, which has caused so much concern for environmental health in the last few decades, give toxicologists an added motivation to push this research further.

6.2.3 The Intestines

The third barrier, and the one covering the largest area of the body in contact with the external environment, is the digestive system. In this context, the best way to block the entry of nanoparticles into the body is by maintaining the regular functioning of the intestinal transit system, with continuous peristalsis, which leads to the excretion of the vast majority of these particles in the feces. Nevertheless, non-biodegradable nanoparticles have been found to cross the intestine wall and introduce themselves into the so-called entero-hepatic portal circulation system*. This circuit carries the particles to the liver via the local portal circulatory system*, and here they accumulate in the Kupffer cells*, a hepatic variety of macrophages that function like other such immune cells, absorbing and breaking down elements identified as 'non-self'. In this case, it takes about five days to eliminate these particles in the feces, but biopersistent particles that are not biodegradable tend to be transported to and then stocked in the mesenteric lymph nodes where they can remain for much longer periods of time.

Overall, experimental research suggests that nanoparticles can cross the three main barriers protecting our 'self' against aggressive and destructive foreign bodies in the nano-scale range, whether they enter by themselves or loaded in macrophages. The disruption and increased entropy such nanoparticles can introduce into the local environment pose a potential threat to our health, and given the different means of entry, this threat is faced by every organ in the body.

6.3 THE BLOOD-BRAIN BARRIER: AN INNER FRONTIER

One of the best protected organs in the human body is the brain. The brain-blood barrier separates it from the rest of the body, keeping it free from a wide range of toxic or infectious material. This barrier is particularly difficult for foreign objects to cross, and yet nanoparticles have been shown to do so, presenting a very worrying threat to human health, given the sensitivity and crucial function of the brain and the central nervous system in general. Using nanoparticles labeled with radionuclides, researchers have been able to trace this passage, suggesting that it takes place across the ethmoid bone located at the base of the nasal pyramid. This would mean that the particles enter the brain like the *Neisseria meningitidis* bacterium (a bacterium provoking meningitis), a risk that is heightened by the local inflammation of the nasal epithelium due to influenza or common forms of allergy. Despite the general effectiveness of the blood-brain barrier in keeping dangerous material away from the brain, it is also equipped with cells specialized in the collection and degradation of any foreign intruders, namely, the local macrophages known as microglial* cells.

6.4 THE DISTRIBUTION OF NANOPARTICLES IN THE BODY

Once the nanoparticles enter the bloodstream, whether loaded in macrophages or circulating independently, they are distributed throughout the body, reaching all the organs, in particular the brain, and the intra-thoracic or intra-abdominal organs. This phase is called 'distribution' and the favored organs for the distribution of the nanoparticles are the spleen and liver, with these two organs receiving as much as 90% of any given dose. Thanks to this mode of transport, nanoparticles are found to build up in various organs including the liver, kidney, spleen, lymph nodes, and bone marrow, organs belonging to the mononuclear phagocytic system and lymphatic system. The accumulation of nanoparticles in these organs can lead to organ damage, and if the nanoparticles are biopersistent (that is to say they are not biodegradable), they extend their stay in the body along with any associated health problems. The US Environmental Protection Agency (EPA) calculated that the inhalation of nanoparticles might cause as many as 60,000 deaths per year, mainly due to the ultrafine particles found in diesel exhaust.

6.4.1 The Metabolism of Nanoparticles

The metabolism—a word derived from the ancient Greek word for change or transformation—of nanoparticles partially corresponds to the thoroughly studied 'classical' metabolic pathway of pharmaceuticals as we will describe it in what follows. Nanoparticles composed of the elements that predominate in organic compounds, carbon, hydrogen, oxygen and nitrogen, are usually broken down faster than those containing silicon, cadmium, gold, silver or other metals not typically found in organic compounds, as well as the rare transition metals*. Once broken down, the soluble metabolic products are secreted and excreted mainly in urine and feces. The biopersistency of a foreign substance (a xenobiotic compound*, which can itself be organic or inorganic) in an organism is evaluated by determining its half-life*, defined as the time (expressed in hours or days) necessary to clear 50% of the material from the organism. For example, the half-life of morphine in the human body is approximately 10 hours. Nanoparticles display half-lives ranging from about one hour to more than 100 days, as has been shown with certain biopersistent ferrite or gold nanoparticles. In case of a long half-life, the nanoparticles are stocked somewhere: either in the organs or in a virtual compartment and the risk of toxicity increases. The dynamics of this kind of exposure has been studied and modeled by specialists in the field of toxicokinetics*.

6.4.2 The Elimination of Nanoparticles

Soluble components of nanoparticles, as well as entire nanoparticles, can be eliminated in the urine, with many of these smaller elements passing through the kidney's filtration system. Nevertheless, the major route of elimination is via the liver with the nanoparticles being excreted in the feces.

6.5 CELL RESPONSE

6.5.1 How Nanoparticles Enter the Cells?

The human body is constituted of approximately 200 types of cells organized in specialized cells or tissues, like retina, pneumocytes, osteoblasts, astrocytes, hepatocytes or macrophages. Nanoparticles can potentially enter virtually all these kinds of cells not only within macrophages, whose function is the uptake of substances identified as 'non-self' and so a potential threat, but as independent species as well. They have principally two means of entry depending on the type of the cell and the nanoparticle's own physico-chemical characteristics.

The largest nanoparticles, like aggregates or microparticles, enter cells principally following phagocytosis, the process of being 'consumed' by macrophages, with polynuclear neutrophils behaving as scavengers responding to receptors on the surface of these particles. In phagocytosis, the membrane of the phagocyte cell engulfs the microparticle and internalizes it in a so called 'phagosome'—meaning literally 'eating body'—that will fuse with lysosomes, smaller vacuoles* containing various enzymes* that are able to break down all the degradable macromolecules of micro or nanoparticles by enzymic hydrolysis*. This is the main role of these 'scavenger' cells in eliminating foreign particles. The other mechanism, shared by various cell lines and tissues, involves the smallest nanoparticles as well as other particles presenting specific chemical properties and is called pinocytosis from the ancient Greek verb '*pinein*' meaning to drink, because this process is devoted to smaller particles. It was discovered after phagocytosis following intensive research using electron microscopy. Pinocytosis is a polymorph mechanism that needs energy from the cell obtained from adenosine triphosphate (ATP*), a universal cellular energy source. On the basis of the study of certain cellular proteins surrounding nanoparticles when they interact with cells, biologists have distinguished four molecular mechanisms: (i) clathrin*-dependent endocytosis*, (ii) caveolin*-mediated endocytosis, (iii) clathrin and caveolin-independent endocytosis and (iv) macropinocytosis. Endocytosis refers to a nano-mechanism that apparently allows nanoparticles' entry into the cell thanks to proteins called clathrins, while caveolins are other proteins found on the external surface of cells: both of them internalize small nanoparticles through an energy-dependent mechanism. In contrast, larger particles, with a size ranging from several hundreds of nanometers to a few micrometers enter cells by macropinocytosis and phagocytosis.

This knowledge of how particles can cross barriers and penetrate cells is of paramount importance not only for explaining the toxic (as well as pharmacological) properties of nanoparticles but also for trying to increase their capacity to cross the cell barrier when researchers are aiming to harness them as nanopharmaceuticals, a concept developed in Chapter 2. Indeed, it has been shown that the concentration of nanoparticles can be up to 100 times higher in cancerous cells when compared to their non-cancerous counterparts due to higher vascularization and a higher degree of retention in interstitial tissue. In addition, cancer cells have an increased rate of nanoparticle uptake compared to normal tissue due to increased metabolic activity. Their higher growth rate demands increased amounts of nutriments, and so nanoparticles are taken in from the bloodstream along with other nutritive substances. As we have already suggested, what is a desirable physiological property in one context, namely, the pharmacological action of a drug in cancer treatment against cancerous cells, is likely to give rise to an adverse effect in non-cancerous cells that are targeted by the same mechanisms because they are dividing rapidly. Thus, what is a therapeutic effect in one context becomes an undesirable toxicological effect in another, making it clear that nanoparticle uptake needs to be evaluated in different human cell lines to give as complete a toxicological assessment as possible.

6.5.2 Cellular Reactions to Nanoparticles

The main response of cells following exposure to nanoparticles is an increase in the production of Reactive Oxygen Species (ROS*). These ROS include mainly superoxide*, the hydroxyl radical* and hydrogen peroxide*. They are produced either by specialized cells like macrophages and polynuclear neutrophils for which destroying 'invaders' (nanoparticles or microorganisms) is an important function, or they can be metabolic products of the nanoparticles themselves, as with metal oxides or fullerenes. Producing large amounts of these ROS* means the end of the cells that produce them, but in general, the best interpretation of this overall mechanism seems to be that this local sacrifice serves to protect the organism as a whole.

The ROS produced in the cell cytoplasm act on chemical sensors to activate a number of genes via a transcriptional protein system called NfκB*. This system activates the transcription* of numerous genes involved in interferon* or cytokine (interleukin*) response as well as kinases*, matrix protease* and proinflammatory proteins* depending on the cell type. This pathway has been demonstrated by the in vitro testing of biopersistent nanoparticles on cells. The final action of such nanoparticles is the activation of certain genes involved in the defense mechanisms of cells through a cascade of cytoplasmic reactions involving specialized proteins and enzymes.

However, if not balanced by antioxidants, this ROS production can lead to cell destruction or at least impairment of vital cell functions by, for example, lipid peroxidation*, and the production of chloramines* on proteins and their denaturation. The ultimate target of the oxidizing capacity of these ROS can be the four nucleotides* of DNA, a phenomenon that can be held in check by 8-oxo-guanosine production, making this compound a useful marker for measuring exposure to genotoxicity. Thus, the main result of the activation of a proinflammatory pathway is the local attraction of polymorphonuclear neutrophils and monocytes from the blood stream: the latter differentiating into fully active macrophages. In certain cases, a persistent inflammation can lead to the cancerous transformation of cells, a process well demonstrated with micrometric-sized particles like asbestos. This outcome has prompted toxicologists to study the genotoxic potential of man-made nanoparticles particularly closely.

In certain cases, when the defense capability of cells is exceeded, the cell dies either by apoptosis* or by autophagy* (Eidi et al., 2012). Apoptosis is a gene-driven cell death mechanism that is launched in various physiological or pathological conditions, such as reduction in cell number, genotoxicity or exposure to ROS. It generally starts when cell fate is uncertain following exposure to a mutagen agent that has damaged the cell DNA. In this case a dividing cell 'prefers' to die instead of transmitting damaged DNA to daughter cells. Here, there are some worrying indications from toxicological research into nanoparticles. Many such particles, including carbon nanotubes and magnetic molecules, have been shown to increase apoptosis levels at high concentrations, both in vitro and in vivo.

Recent research has shown that cells have other mechanisms for purging themselves that fall short of the cell death triggered by apoptosis*, and that these mechanisms can also be triggered by nanoparticles. Autophagy* is another form of cell auto-destruction that was first brought to light in the 2010s, following the exposure of immune cells to viruses, which here we can take to be a kind of naturally occurring nanoparticle. Autophagy is a mechanism that allows cells to recycle organelles like mitochondria*, which in this case would be the more specific mechanism of mitophagy*. This process depends on a gene-driven program that commands a cell to destroy its own mitochondria in autolytic vacuoles. This phenomenon can be observed when macrophages are exposed to polymeric* nanoparticles that enter the mitochondria, impairing essential cell functions in these organelles that provide the energy needed to maintain the cell in activity. Here, unlike the cases of apoptosis or necrosis, only a part of the cell is destroyed.

6.6 THE PARADOX OF THE DOSE–FUNCTION RELATIONSHIP WITH NANOPARTICLES

As with any toxicological model, it is important to evaluate the potential toxic effect of exposure to any nanoparticle in relation to the dose received by the organism. In most but not all of these models in toxicology, increased dose means increased toxicity. But this relationship between dose and response is generally more complex in the case of nanoparticles, with smaller doses being particularly effective in stimulating physiological function, including the toxic ones outlined earlier, while higher doses generally induce less reaction. This biphasic effect, more stimulation at lower doses and less at higher concentrations, is called hormesis* and is seen in a range of toxicological effects associated with nanoparticles, especially those that concern cell viability, like apoptosis.

6.6.1 New Paradigms for the Toxicological Testing of Nanoparticles

'Nanotoxicology' therefore is a very particular kind of toxicology. Due to specific properties and mechanisms of action associated with nanoparticles, toxicologists have had to change certain of their previously established paradigms for toxicology.

6.7 HOW TO EXPRESS DOSE?

First of all, the notion of dose has evolved. The dose of toxicants or pharmacological substances was classically expressed in terms of mass per volume of unit doses, for example, in micrograms per milliliter, a measure of concentration independent of the living system or subsystem under consideration: cells, organs or the whole organism, although for an individual, dose is expressed in terms of mass of the substance per unit body mass (kg) of the subject. Using this kind of dose expression for the same nanoparticle (but present in different forms), it is impossible to compare the activity of nanoparticles of different sizes with respect to the same endpoint. For example, we cannot use these classical measures to express the polynuclear neutrophil* content of bronchoalveolar lavage in the lungs of rats exposed by inhalation to TiO_2 as a nanoparticle of two different sizes, 20 and 250 nm. Two dose–effect curves are needed to present the effect of these two nanoparticles as a function of a dose expressed in μg/kg: one for rodents exposed to an aerosol of 20 nm particles and another one for those exposed to one of 250 nm particles. Once the dose is expressed in cm^2 of nanoparticle area over body weight or per exposed surface, however, the two experiments can be merged into a single dose–effect curve, thereby resolving this apparent complexity. The effect of a nanoparticle is not a consequence of the exposure to a specific mass of the nanoparticle but a consequence of exposure to its unfolded surface. The surface plays a mechanistic role that allows us to understand the nanoparticle's effect in terms of its capacity for interaction with living systems. Here, we can recall that nanoparticles are generally particles with an exposed, reactive surface area higher than 60 m^2/g. Thus, they are able to interact via this extensive surface area to block or inversely to activate living systems, a property already predicted by Richard Feynman* in his famous talk 'There's Plenty of Room at The Bottom' (Feynman, 1961). The nanoscale is a synonym for high surface area, and nanoparticles react with living systems through surface contact: the higher their specific surface area the greater their reactivity. This contrasts with the more specific, chemical action of traditional pharmacologically active molecules, which react with given receptors or enzymes to induce their effects on living systems. These active molecules follow biochemical laws like the affinity of a ligand for a receptor or that of a substrate for an enzyme. From this perspective, a virus can be thought of as a biological nanodevice or as an inert nanoparticle when outside the cell, as a virus does not have the capacity to 'live' independently of its integration into a host organism. The virus can be thought of as coming to life once inside the infected cells by hijacking their metabolism. In order to get inside, the virus passes through a specific lock (a receptor) thanks to its 'keys', which populate its outer surface, having been acquired during its long-term co-evolution with humans and other animals.

6.8 IMMUNE PARALYSIS BY OVERLOADING

Another novelty introduced into the evaluation of the dangers of nanoparticles is a phenomenon called immune-cell 'paralysis'. Moss and Wong have described the kinetics of this phenomenon related to the uptake of nanoparticles, where a single macrophage is able to integrate around 10^6 nanoparticles of TiO_2 (about 20 nm in size) in less than five minutes (Moss & Wong, 2006). Thus, following exposure to this nanoparticle, the macrophage becomes saturated and so can no longer absorb any such nanoparticles from the medium. The maximum loading capacity of a macrophage has been determined to be over 1.5 million for a 20 nm-size particle, almost 10,000 for one of 250 nm, and less than a hundred for a 3,000 nm (3 μm) particle (1,540,000, 9,900 and 69). These numbers correspond to approximately 6% of the volume of a macrophage, indicating that above this value, the macrophage becomes overloaded, no longer capable of biological reactivity, and so becomes 'paralyzed'. In classical toxicology, reaction equilibrium is driven by laws such as mass action, following the relationship described by the Michaelis–Menten equation, affinity and other biochemical laws; in the case of enzymes* for example, they continue to synthesize the substrate into the specific product as long as the substance is present.[4]

[4] Provided it is not down-regulated by a feedback loop.

Nevertheless, when we deal with nanoparticles, we have to take this notion of immune-cell paralysis due to overloading into account.

6.9 BIOPERSISTENCY* AND 'FRUSTRATED PHAGOCYTOSIS'

Living organisms and their tissues are mainly composed of non-metal elements (carbon, oxygen, hydrogen, nitrogen and phosphorus) with only traces of titanium, silicon, cobalt, silver, etc. being found distributed throughout the body, so when nanoparticles contain such transition metals* or other metals not usually encountered in living tissue, we have to explore their toxicity more thoroughly. Special emphasis has to be placed on the issue of biopersistency*, as these elements are rarely biodegradable in contrast to more or less soluble substances (xenobiotics*), which are metabolized by a battery of enzymes, especially those involved in detoxification. Biopersistency is an approach, or maybe even a paradigm developed in the 1990s to evaluate the danger of asbestos exposure whose long (>20 μm) fibers can persist in the lungs for years, inducing a kind of paralysis of macrophages and the formation of giant cells—composed of macrophages that merge together—that favor chronic inflammation very deep in the lung. This phenomenon has been termed 'frustrated phagocytosis', as these materials are not biodegradable and continuously induce the influx of macrophages deep into the lung, producing lung fibrosis in a time-dependent manner, and provoking cancers like mesothelioma, a cancer of the lining of the lung. Frustrated phagocytosis and giant cells, so called syncytium*, have been observed in rats following the inhalation of carbon nanotubes, providing they have a length to diameter ratio greater than 3. Thus, researchers need to be very attentive to this kind of giant morphology of macrophages that can be observed in toxicological studies involving human exposure to nano and even microsized particles. Here is another aspect of nanotoxicology that does not correspond to the model of classical toxicology, where such problems with macrophages are rarely observed.

6.10 THE CORONA OF NANOPARTICLES

In the early 2000s, researchers showed that once in biological media, nanoparticles have the property of attracting proteins from serum due to the high surface to volume ratio of nanoparticles and their marked ability to interact with the environment. As various proteins have the potential to stick to different surfaces to generate biofilms, it is not surprising to find proteins forming a crown around spherical nanoparticles. The formation of a 'corona' follows protein adhesion to the surface of nanoparticles due to hydrophobic ionic and Van der Waal's bonds. There are about 300,000 different proteins in human blood serum, and proteins are known to adhere to virtually all kinds of surfaces whether polymeric, metallic or featuring transition metals. Thus, once in the body, each species of nanoparticle has its own fingerprint of corona proteins that confers it a special fate in the living organism, although the result is unpredictable without any screening or detection using mass spectrometry*. Metallic nanoparticles have been shown to adsorb proteins involved in coagulation, although this does not compromise the coagulation function of blood. Certain polymeric nanoparticles retrieve cell growth factor from human serum, conferring better viability or even a differentiation potential to exposed cells. Their extensive study will allow researchers to better target cells using synthetic nanoparticles, developing theranostic tools as outlined in a previous chapter. The influence of the so-called coronome*[5] associated with nanoparticles is largely underestimated in nanotoxicology, although the interaction between nanoparticles and biological fluids has been the subject of recent research. As more is known about these coronomes, their individuality is becoming more evident, and the signature of these coronomes has increasingly been a focus for studies. Thus, the characterization of 'coronomes' is a new research field in nanotoxicology that will allow researchers to better understand interactions between nanoparticles and plasma proteins and the modification of their initial (pre-corona) profile compared to the one acquired following contact with serum proteins. In terms of the application of this phenomenon in nanomedicine, once the coronome has been extensively studied, nanoparticles will without doubt be valuable tools for locating and even for neutralizing proteins of clinical interest.

6.11 ARE STEALTHY NANOPARTICLES A MYTH?

As we saw in the chapter on nanopharmaceuticals, researchers in vectors and encapsulation have been aiming to synthesize nanoparticles capable of escaping detection by the immune system, principally the mononuclear phagocytic system*, with its main actors the macrophages and dendritic cells*, which recognize the 'non-self' with a high degree of accuracy using a battery of receptors, usually from the *Toll Like Receptor** family. Today, scientists in search of long-circulating nanoparticles are looking in detail at the behavior of macrophages as they are involved in the 'cat-and-mouse' game of eliminating foreign materials. The goal is to exploit this knowledge to allow nanoparticles to be able to circulate longer in the blood stream, thereby increasing the probability of their reaching their targets. Synthesized and biodegradable nanoparticles rarely have a half-life in biological fluid of more than a day due to their rapid uptake by the mononuclear phagocytic system. Polyethylene glycol (PEG)-covered gold nanoparticles are longer lasting, and they have led to some clinical applications in oncology as we saw in a previous chapter. Nevertheless, they also elicit immunity, evidenced by antibody anti-PEG synthesis, suggesting that such nanoparticles are not as stealthy as was initially hoped. Other avenues of research are being explored to increase the bioavailability of nanoparticles for target cells, while reducing their exposure to macrophages. Synthetic hydrophilic* and zwitterionic* materials have been proposed as replacements for PEG to customize the surface of vectors, and some researchers even imagine mimicking the plasma membranes of red blood cells, all in an attempt to prolong the half-life of these nanoparticles in living organisms. Nevertheless, the initial optimistic claims made for the potential of this approach have probably been overstated.

There is another problem with the use of nanoparticles in drug delivery; if the nanoparticles are not retrieved by the mononuclear phagocytic system, they can be taken up by other non-target cells like neurons, cardiac or renal cells where they may be potentially harmful. Once again, it is very important to test these

[5] A subset of serum proteins that are bound in the corona of a nanoparticle.

nanopharmaceuticals from a 'nanotoxicology' point of view, looking at their potential uptake by all the main tissues, offering another area of application for the organs-on-chips that we will be considering in a subsequent chapter.

This non-exhaustive inventory of areas of interest in nanotoxicology clearly indicates the need to adopt and develop new paradigms in the field. This means developing structural alerts (both in silico and in vivo) for nanoparticle toxicity, just as they are being developed for other potentially toxic chemicals or pharmacological molecules. The field also clearly needs to be developed in terms of fundamental research into this specialist area of toxicology. The development of nanotoxicology, however, is taking place in a general context of limited and often underfunded research. There is a marked imbalance in studies in this field, as it is estimated that the ratio of toxicologists to chemists or engineers working on new materials is somewhere near 1/1,000 meaning (very approximately) that for every 1,000 new nanomaterials that are synthesized, only one is evaluated for its potential dangers or its global risk. The relative paucity of nanotoxicology is also due to the substantial difference in time scales needed for risk assessment when compared to the rapidity and profligacy of the dominant forms of chemical synthesis. This being said, it is not clear that toxicological studies can keep up with the production of synthetic organic chemistry either, and so this imbalance of both research funding and effective capacity is not a problem specific to nanotoxicology.

In the absence of any clearly established threat to human health from nanoparticles, we have presented the elements of what distinguishes toxicology in this domain from more fully explored areas, in particular the field of toxicology that studies the effects of microparticles. Thinking about the pathways and potential for damaging the living organism that is opened up by this change in scale is illuminating, although not particularly reassuring. This being said, while the threats posed by nanoparticles to human health seem potentially significant, they remain for the moment largely hypothetical, but this toxicological research needs to continue. Given the novel mechanisms that have already been observed and explained, such as the overload of macrophages, perhaps the brightest sign in this context is the growing interest that a new generation of researchers is taking in this emergent field of nanotoxicology. It is some comfort to know that more and more scientists will be looking into these complex questions, motivated by an interest in the health implications of artificial (and natural) nanoparticles.

Following the logic of this book, we have been thinking about toxicology in the context of nanomedicine. The wider questions about the potential health threats of nanoparticles take us quite far from the central concerns of this domain, which are intervention with the goal of curing disease, repairing bodies or improving human health. But the two are clearly interconnected. It is not just coincidence that the same properties of nanoparticles that allow them to target infectious or neoplastic disease more effectively than their larger counterparts should also expose the body to greater potential risk if similar particles are introduced into the body for the same or for any other reason. It is precisely this kind of trade-off between the benefits of using nanoparticles for health purposes and the risks of harm associated with their use that prompted us to frame this book with the figures of Panacea and Pandora's box. In a more general treatment of the issues around nanoparticles and nanomaterials, this question would have to be pushed further. Finding the right balance between the interests of innovation in the domain of the nano versus the potential, but unfortunately not entirely predictable health hazards associated with such new materials is as much, if not more a political and a social question than a scientific one, but it is one that should be addressed in as clear terms as possible.

6.12 TO GO FURTHER

Other useful references include a general review of the biomedical use of nanoparticles (Rihn, 2020) a reference work in toxicology (Bernstein, 2007) an overview of nanosafety (Lebre et al., 2022), and general texts on the interactions between nanomaterials and organisms (Sutariya & Pathak, 2014; Staroń et al., 2020). Reviews of the toxic effects of nanomaterials (Khan and Arif, n.d.; Saquib et al., 2018) are complemented by a more specific study of silicon and titanium oxides (Peters et al., 2020).

Bernstein, D. M. (2007). Synthetic vitreous fibers: A review toxicology, epidemiology and regulations. *Critical Reviews in Toxicology, 37*(10), 839–886. https://doi.org/10.1080/10408440701524592

Khan, H. A., & Arif, I. A. (Eds.). (n.d.). *Toxic effects of nanomaterials* (H. A. Khan & I. A. Arif, Eds.). Bentham Science Publishers. ISBN: 9781608052837, 147 pp.

Lebre, F., Chatterjee, N., Costa, S., Fernández-de-Gortari, E., Lopes, C., Meneses, J., Ortiz, L., Ribeiro, A. R., Vilas-Boas, V., & Alfaro-Moreno, E. (2022). Nanosafety: An evolving concept to bring the safest possible nanomaterials to society and environment. *Nanomaterials (Basel, Switzerland), 12*(11), 1810. https://doi.org/10.3390/nano12111810

Peters, R. J. B., Oomen, A. G., van Bemmel, G., van Vliet, L., Undas, A. K., Munniks, S., Bleys, R. L. A. W., Tromp, P. C., Brand, W., & van der Lee, M. (2020). Silicon dioxide and titanium dioxide particles found in human tissues. *Nanotoxicology, 14*(3), 420–432. https://doi.org/10.1080/17435390.2020.1718232

Rihn, B. (Ed.). (2020). *Biomedical application of nanoparticles*. Taylor & Francis.

Saquib, Q., Faisal, M., Al-Khedhairy, A. A., & Alatar, A. A. (Eds.). (2018). *Cellular and molecular toxicology of nanoparticles*. Springer.

Staroń, A., Długosz, O., Pulit-Prociak, J., & Banach, M. (2020). Analysis of the Exposure of Organisms to the Action of Nanomaterials. *Materials (Basel, Switzerland), 13*(2), E349. https://doi.org/10.3390/ma13020349

Sutariya, V. B., & Pathak, Y. (Eds.). (2014). *Biointeractions of nanomaterials*. CRC Press.

6.13 REFERENCES

Eidi, H., Joubert, O., Némos, C., Grandemange, S., Mograbi, B., Foliguet, B., Tournebize, J., Maincent, P., Le Faou, A., Aboukhamis, I., & Rihn, B. H. (2012). Drug delivery by polymeric nanoparticles induces autophagy in macrophages. *International Journal of Pharmaceutics, 422*(1–2), 495–503. https://doi.org/10.1016/j.ijpharm.2011.11.020

Feynman, R. P. (1961). There's plenty of room at the bottom. In H. D. Gilbert (Ed.), *Miniaturization* (pp. 282–296). Reinhold.

Moss, O. R., & Wong, V. A. (2006). When nanoparticles get in the way: Impact of projected area on in vivo and in vitro macrophage function. *Inhalation Toxicology, 18*(10), 711–716. https://doi.org/10.1080/08958370600747770

7 Organs on Chips, Miniaturization and Medical Specialties

The Different Logics of Nanomedicine

In this penultimate chapter before drawing our conclusions, we want already to take a step back and think about nanomedicine from a different perspective, to try to make sense of the apparent slowdown in the growth of the field over the last few years.

To do this, we will consider a more general trend toward miniaturization in medical science and technology that clearly mirrors a much wider technological movement affecting every aspect of contemporary science, industry and, we are tempted to say, life. From computers to telephones to motorized personal transport, everything is getting smaller, but this reduction in size does not necessarily imply an ultimate transition to the nanoscale.

How can we best understand this relationship between miniaturization and the nano? To try and answer this question, we propose the following approach: first, we analyze certain techniques—in particular the 'organ-on-a-chip'—that are motivated by miniaturization (among other things), working to produce ever more compact apparatus. Although these techniques now mobilize certain nanotechnologies, they do not necessarily have the ambition of arriving at the nanoscale. This should be clearer when we have presented some examples of organs on chips. We then take advantage of these examples to consider the limits to miniaturization in more general terms. While certain medical techniques seem destined to continue down the path of miniaturization until they arrive at their ultimate molecular incarnation, it is not going to be the case for all of them, and the revolutionary potential of certain approaches that are often prematurely classed as nanosciences might well lie elsewhere.

We start, then, with the example of the organ-on-a-chip because it is often included in discussions of nanomedicine, while its status as a nano-object is, as we shall see, hard to justify. Thus, the organ-on-a-chip in its various forms allows us to consider the different kinds of motivation that lie behind a more general movement toward identifying and retaining essential complexity in ever smaller devices. At the same time, we will think about the limits to this miniaturization, as there clearly are limits, and thinking about these limits inevitably takes us back to the domain of the nano, as the ultimate horizon for any such ambitions. With this miniaturization trend in mind,

DOI: 10.1201/9781003367833-7

we then want to return to a global consideration of nanomedicine, and think about it not as a medical discipline in and of itself, but as an approach that can function across the whole spectrum of contemporary medicine, contributing to many if not all of today's medical specialties. We will consider a few such areas of specialization; oncology, transplant medicine, reproductive medicine and nephrology, in order to see what lessons we can learn from the dynamics that nanomedicine has introduced across the different medical specialties. In the end, our initial and fundamental error might well have been to think of nanomedicine as a new form of medicine. It could turn out to be more appropriate to conceive nanomedicine as a new approach that will ultimately be distributed across medicine in general. Thus, it seems probable that the mature destiny of nanomedicine is to disappear as a clearly identified medical field, and instead be integrated as a series of sub-fields or a nano-contribution to each medical specialty considered in turn. Before embarking on these general considerations concerning the future of nanomedicine as a distributed sub-specialty, however, let us look at the organ-on-a-chip and use it to explore the key relationship between miniaturization and the nano.

7.1 THE GOALS AND LIMITS OF MINIATURIZATION

In other chapters of this book, we have had the occasion to evoke early images of the nanosciences—a fantastic Drexlerian vision of nanofactories populated by nanorobots—tiny machines capable of constructing, dismantling or rearranging molecules atom by atom. This image was inspired by the prowess of the scanning tunneling electron microscope (STEM), a potential only confirmed by subsequent atomic force microscopes. Already in 1989, Donald Eigler and Erhard Schweizer, working for IBM in California, were able to use an STEM to write the three initials of their company by placing 35 xenon atoms on a nickel crystal. This manipulation of individual atoms, unimaginable even just a few decades earlier, placed the nanosciences on a par with science fiction, and inspired nano-visionaries, most famously Eric Drexler, to imagine the production of customized molecules atom by atom. The image of nanorobots as tiny molecular-sized machines capable of carrying out practically any task enjoyed a certain fashion and some people continue to see the nano in terms of these kinds of manufacturing robots. Nevertheless, there is a point at which miniaturization is no longer possible, or at least not on the same terms, because when you arrive at the level of molecules and atoms, everything changes. If we think of model railways, where model builders have reduced railway trains by different scales (O scale trains are models of full-size trains reduced by a factor of 45, HO by 87, and Z scale ones are reduced by a factor of 220), we might imagine we can go on doing this indefinitely. While for Aristotle it was conceivable to divide matter without any limits, following the triumph of modern atomic theory in the nineteenth century, chemists and physicists have had to recognize the physical limits of such reduction of size. There comes a moment when you can no longer divide material and expect it to behave the same. Today, scientists no longer consider the atom to be as indivisible as its name would suggest; the atom consists of different sub-atomic particles, themselves made up of the fundamental constituents of matter.

Even before reaching the stage of such sub-atomic particles, when one arrives at the frontiers of this molecular or atomic scale—the nanoscale—the rules governing what happens when one divides matter change fundamentally. Dividing an oxygen atom in half (and we cannot, in the spirit of the original Greek atomists, think of it as cutting a small ball in two) is not a mechanical manipulation and would not give two mini oxygen atoms, but rather might give us two beryllium atoms (the beryllium atom possesses half as many protons as an oxygen atom). Thus, the properties of the products of this division are completely different from the original atom. Even before we get down to the atomic scale, however, dividing matter can produce new and even unexpected properties. These properties can provide opportunities for nanoengineers, just as they can introduce new threats to human health, something we have already looked at from various angles in the chapter on nanotoxicology. With the development of working units at the molecular level, molecular nanorobots are already being developed to perform primitive mechanical operations at the nano level. But what about living organisms and their constituent organs? Are miniaturized models of human organs inevitably going to take us down to the nano level? Thus, for example, we can ask whether the nano ought to serve as the ultimate scale for the organs on chips (labs on chips and tumors on chips) we will be looking at in this chapter?

7.2 MINIATURIZING ORGANISMS AND ORGANS. THE POSSIBILITIES AND THE LIMITS

The liver is an exceptional organ in terms of its capacity to function even when severely damaged. Many people live almost normal lives with 70% or even only half their healthy liver function. And people live more or less normal lives with just one kidney or one single lung. But what about a heart half the diameter of a normal adult size? Such a small human heart would be incapable of generating enough force to maintain an adequate flow of blood through the body to keep the adult in question alive. But wouldn't it be possible to compensate for this lack of blood pressure with a proportionally reduced body size? There are small or 'pygmy' versions of a number of animals, like goats, but these smaller breeds have different proportions between their body parts when compared to their larger relatives. You cannot just reduce a living being in size while keeping all the proportions the same like you can with model railways. Indeed, if the model locomotives were supposed to work the same way as the full-scale originals, their proportions would likewise have to be altered. Last, but not least, in animals, we have to consider their cells, the constitutive units of any animal or plant. While cell size does vary, smaller animals are generally composed of fewer cells than larger animals, and past a certain limit cannot sustain the same complexity and diversity of function that rely on differentiated tissues and the system of organs in the body.

Thus, as any amateur of Bonsai knows, while it is possible to miniaturize certain plants or animals, there are limits. Functional living cells and the collection of specialized cells into organs impose certain constraints to this reduction in scale. This translates into a double limitation for the organs on a chip we are about to look at.

While the aim of this technology is to have cells functioning in similar conditions to those in the original living organ, the chips cannot provide the same biological context as the cells enjoy in situ in a living organism. On the other side of the equation, the cells transplanted into an organ-on-a-chip impose a lower limit to its size that is well above the nano-scale; one cannot just divide cells down to the molecular or atomic level and expect them to continue performing their multiple functions as cells. Thus, despite the feat of reducing organs to the level of cells, we remain far from the nano-scale: human cells are around 15 micrometers (μm) in diameter, and although red blood cells are under 7.3 μm wide, even this microscopic size is well outside the generally accepted range for nanoparticles.

One response to these observations is simply to declare the organ-on-a-chip as definitively non-nano and hence rule it out of any discussions of nanomedicine. But there is an alternative approach; although it is quite natural to think of nanoscience as starting from the nanoscale (on the level of atoms and molecules), because this is what is most remarkable about it, one can also think of it in the other sense—in terms of miniaturization. The error made by Eric Drexler and other early 'nanoenthusiasts' was to think that everything was miniaturizable down to the molecular scale, following the kind of approach seen in the model railways we were thinking about earlier. While Drexler and his acolytes were particularly interested in robots and the possibility of miniaturizing machines down to this scale, which, despite the considerable challenges it faces, remains an active project today, the problem is much more acute when it comes to living tissues and cells.

Commentators on nanotechnology often evoke Richard Fleischer's 1966 film, *The Fantastic Voyage* as a vision of what the nano might one day be able to do—send miniature machines into the body to treat disease. In the film, the scientist responsible for the miniaturization technology has a blood clot in the brain and the American government shrinks a submarine—including its full crew—down to the size of a red-blood cell to be able to remove the obstruction. While an entertaining adventure, the premise of the film is pure fantasy. To think that you could reduce a human being to a size smaller than a blood cell is simply unrealistic—as we will see with the organ-on-a-chip, cells are not greatly reducible. While there is a huge difference in scale between a whale and an ant, for example, their integrated functioning does not allow one to be reduced down to the size of the other like a model railway train. This being said, despite their difference in size, both organisms do execute similar physiological functions such as motion, digestion and reproduction, and share many protein structures and metabolic reactions in common. This is one of the paradoxes of miniaturization: the possibility of having smaller versions of functioning living organisms and the impossibility of just shrinking them down in size without limit, and it will help us to think about a core feature of the nanosciences. Thus, while nanomedicine interpreted in its strictest sense takes us to the limits of the molecular and beyond toward the atomic, any movement toward miniaturization can, in a sense, be considered as part of the same approach. Here is the real interest of the organs on chips we will be looking at in what follows, but even before turning to the organ-on-a-chip, to better situate this innovation it will be helpful to consider its precursor, the laboratory or lab on a chip.

7.3 FROM THE LAB(ORATORY)-ON-A-CHIP TO THE ORGAN-ON-A-CHIP

The miniaturization of techniques used in experimental biology is not a new phenomenon; starting at least in the 1970s, all the principal techniques were reduced in size before becoming computer-controlled during the 1980s and 1990s. These miniaturized techniques included chromatography, electrophoresis and ultracentrifugation among others. The leader in miniaturization in recent biology, a phenomenon accompanied by a stunning increase in capacity, is of course genetic sequencing. Genetic sequencing tasks that used to take months, now take days or even less, and sequencers that used to fill large rooms now fit easily on the laboratory bench. Around 2000, the term 'chip' came into popular use for qualifying the screening of biological macromolecules (DNA, RNA or proteins) using micrometric spots of thousands of DNA oligonucleotides* produced by in situ synthesis and laid out in these systems as we have already seen in the chapter on microarrays. The production of these chips generally relies on printing using photolithography, a technique based on laser technology that was developed during the 1980s by start-up companies like Affymetrix. This approach has allowed precision printing on a micrometric and even a nanometric scale, thanks to the development of microelectronics and robotics. A lot of analytical techniques in molecular biology and chemistry were scaled down during this period and the vast range of possibilities opened up by photolithography* allied to the mastery of microfluidics*—the science and engineering of handling very small volumes of fluids—allowed scientists to develop the first labs-on-chip for analytical purposes. One example of this approach is the use of systems based on antibody or antigen detection that function with a simple microfluidic system driven by capillary force. This technique has been used to produce pregnancy tests based on the revelation of the human chorionic gonadotropin (hCG) hormone*, and more recently COVID-19 tests that use the antibody against the SARS-CoV-2 to detect the presence of the virus. Like the glucometers that use strips that can be charged with a droplet of blood, these test systems are referred to as point-of-care diagnostic platforms, as they have become readily transportable out of the original controlled laboratory setting in which they were developed. These kinds of systems continue to be developed and expanded for routine biological blood or urine analyses in laboratories using micrometric samples, as well as being deployed at the bedside or even sold over the counter for domestic use. A number of companies now use this lab-on-a-chip approach in order to supply point-of-care diagnosis for various diseases, mainly hepatic, cardiac and infectious, by measuring plasma levels of markers using coagulation. The technique can also be used for determining a patient's blood group, or performing hematological or biochemical analyses of samples of human urine, blood, tears or mucosal swabs.

Labs on chips have been deeply dependent on the domain of microfluidics for their development. Theories and techniques derived from the study of microfluids applied to these objects have allowed researchers to generate gradients of substances or drugs of interest on microscopically small samples of tissue. By using extracellular membranes, this approach has allowed diffusion studies in three-dimensional models. When associated with mechanical sensors, the same tests can be conducted using fewer cells and less cell culture medium. This approach allows organs on chips to reproduce tissue-to-tissue as well as interface (e.g., air–liquid) interactions. In this context, it is possible to mimic the smallest physiological unit of an organ, such as a myocardial fiber.

7.4 THE ORGAN-ON-A-CHIP: IS THE FUTURE OF METABOLIC TESTING ALREADY HERE?

As we have already suggested, organs on chips as physical devices are far from attaining the nano scale, but they remain a remarkable feat of miniature biological engineering. A few centimeters in length, these plastic boxes look like three-dimensional bricks, something like the Lego® blocks so popular with children. Produced in standardized sizes, they come with various points for hooking them up to the microfluid supplies that bring them to life, so to speak. These boxes contain human cells that have been transplanted into an artificial space in order to mimic a human organ: a kidney, blood vessels, potentially any organ in the human body, now placed in the ideal conditions for observation. The principle is straightforward: human cells taken from the organ of interest are cultured and laid out within the box where they can be irrigated with nutritional fluids to preserve their physiological functions. The nutritional flow can then be combined with a drug that researchers want to test on these cells, and the cellular response observed directly within these miniature 'organoids'*. The key to the functioning of such micro-organs is precisely this flow of nutrients (and the drugs under test) through the tubes that make up these simplified and yet functionally sophisticated artificial organs. The idea is to establish an environment that is somewhere between the isolated cells of an organ (passively kept alive and rendered temporarily functional by an appropriate nutrient medium) and the complete functioning organ itself (the whole lung in situ in a model animal, for example). This novel environment also has the advantage of being transparent, both literally and in terms of the measurements of physiological function that can be made on the system. The researcher is no longer limited to noting the input, waiting and then recording the output of this kind of pharmacological experiment, but can observe and measure the functioning of the cells in real time, throughout the test procedure, potentially gaining valuable therapeutic and toxicological information. These models allow researchers to follow the metabolism of a given substance and any subsequent biochemical changes induced in a single cell using confocal microscopy. Of course, all these possibilities generate yet more data. When all the cells are taken directly from the same patient, this information might be quite sensitive, bringing us back to the ethical problems faced by precision medicine as a therapeutic approach, which operates largely with patients' genetic data.

To give the reader an idea of how these organs on chips work, we will work through three examples of physiological systems that are the objects of ongoing research.

7.5 LUNG CELLS

Lung cells have been modeled using an organ-on-a-chip that deploys a unitary alveolus with a flexible polymeric membrane separating epithelial cells from endothelial cells grown on either side. This is considered to be an exemplary model of this kind of miniaturization (combined with the elimination of the animal as experimental subject), as with 'low-tech' tests on cell lines or bacteria. A vacuum applied laterally allows the reproduction of the mechanical stretching associated with breathing, as the tissue would experience it in vivo. Thus, this organ-on-a-chip simulates the mechanical deformation of the alveoli–capillary barrier induced by the motion associated with respiration. This is important, as such barriers respond differently to pollutants—especially nanoparticles—as we explained when treating nanotoxicology in the previous chapter, when they move, and so this kind of organ-on-a-chip device facilitates studies on the physiopathological findings of diseases like pneumonia, pulmonary thrombosis, asthma and more recently, drugs directed against severe acute respiratory syndrome coronavirus (SARS-CoV-2). The simulation of the dynamics of these tissues is vital to modeling the organ in situ.

7.6 HEART CELLS

Just as the movement and stresses of the lung tissue are important to simulating its natural context when transplanted into the organ-on-a-chip, the electrical environment of cardiac cells is critical for reproducing their function and for studying the effects of toxins, hypoxemia-induced myocardial injury, or for studying candidate drugs intended to stimulate or reduce the rate or the force of the heartbeat. These chips are becoming increasingly sophisticated, and when constituting an organ-on-a-chip using cardiomyocytes*, for example, the deformations of polymeric membranes used to simulate the movement of the heart can be supplemented with electrical currents to further imitate the situation of stimulated cardiac cells in vivo. It is hoped that this model of a heart-on-a-chip will allow researchers to avoid using cohorts of dogs to test drug leads or cosmetics. The dog is an animal whose cardiovascular system resembles the human one, leading to the sacrifice of many thousands of these animals in animal experimentation every year.

7.7 SKIN

A third example is provided by the reconstruction of skin using a skin-on-a-chip 3D model with primary dermal fibroblasts* and (epidermal) keratinocytes* microperfused by endothelial coated channels. This kind of system can remain viable for up to two weeks, and it can be used to screen irritant substances in cosmetics as well as to study the healing of wounds. Skin is an area that is generating a lot of research, particularly given the scale of research and development in the cosmetics sector. Vascularized human skin equivalents are just one example of how researchers are improving the skin-on-a-chip model, and there is every hope that these skin-on-a-chip techniques will be capable of fully replacing animal testing for the majority of beauty products that are applied directly to the skin (Risueño et al., 2021).

Numerous pathological conditions involving the brain, skin, liver, kidney or gut can be reproduced using these organs on chips which, as we explained earlier, can be subjected to electrical, mechanical and chemical stressors. The effects of these stressors induce signals that can be recorded by the appropriate sensors, bringing an integrated system mimicking physiological or pathological conditions right to the researcher's laboratory bench.

With commercial labs on chips and organs on chips becoming increasingly available, their contribution to the world of ready-to-use devices will be significant in healthcare, particularly in their initial market, which aims at replacing the use of animals in preclinical studies, thereby economizing costs in this area and reducing animal suffering.

7.8 MULTIPLE ORGANS

So far, we have been considering single organs, a way of testing the effects (whether positive or negative) of pharmaceuticals or cosmetics on just the skin, for example. While aiming to serve this market, developers of organs on chips have already been venturing into more complex models that put different organs together. The liver-and-kidney-on-a-chip is just one example of a multiple organ chip that can serve to illustrate this kind of approach (Theobald et al., 2018). Hepatocytes* can be co-cultured or bioprinted along with the four principal non-parenchymal cells: liver sinusoid*, bile duct, hepatic stellate* and Kupffer cells* to develop a perfused liver chip with vascular layers. This kind of integrated organ paradigm has been shown to respond to oxidative stress, drug or toxicant injury, following either acute or subacute exposure. Using these chips, researchers can track biochemical or biophysical endpoints like mitochondrial* function, oxygen pressure, acidity or cell-secreted markers. In a similar way, nephrotoxicity or neurotoxicity can be screened by measuring dynamic changes in the system and by using perfusion paradigms. These systems allow researchers to stay closer to in vivo experimental conditions and offer many advantages when compared to classical in vitro co-cultures of cells. They also enhance cell sensitivity and response to drug exposure. Last but not least, the diseased cells belonging to the patient who is going to undergo treatment can be used for the drug screening process.

Known toxic substances or drugs with well-characterized side effects are used (alone or in combination) to test these organs on chips to validate their sensitivity before they are deployed for screening new substances, thereby ensuring their pertinence on well-known terrain, before entering into 'terra incognita' as it were. Generally, devices operating with these microfluidic channels are found to be more sensitive and give more realistic results when compared to traditional in vitro screening, increasing the signal-to-noise ratio in their outputs.

These chips can also be used to assess the cytotoxicity and certain parameters—such as the pharmacokinetics* and pharmacodynamics*—of metabolized anti-cancer drugs in interconnected organs like the liver and bone marrow, with artificial blood vessels joining them together.

The flexible physical nature of the materials that make up these boxes means that they can mimic the dilations and contractions associated with the flow of blood through arteries or veins, for example, making the environment even more 'realistic'. Indeed, cardiomyocytes* obtained from mesenchymal stem cells* (MSCs) have been shown to develop the excitation and contraction coupling properties characteristic of heart function. Layered on an electromechanical sensor, these contractions can be recorded following exposure to various activating or inhibiting cardiac drugs. Measures of contractile cardiac force and the flux of Ca^{++} (calcium, an ion that plays an important role in myocardial contraction) have been recorded following exposure to cardiostimulants. In an additional step, these drugs can be metabolized in an interconnected liver on a chip, allowing an initial check for cardiotoxicity. This is just one example of the kind of interconnected experiments that can be performed using these organs on chips, and with the technology still evolving, more and more physiological units representing every organ can be modeled on a microfluidic chip and integrated into a complex yet modular system. The aim of such designs is to reduce and to model the complexity of a human organism in a representative subset of metabolically active tissues in order to define the pharmacodynamic and pharmacokinetic parameters of a given drug.

The idea is to move beyond working with a simple sample of cells from an organ by instantiating the physical context of the organ, rendering the environment as close as possible to that of the cells *in situ*. This is exemplified with bone marrow, one of the largest organs in the human body which continuously maintains and regulates the production of three lineages of the cells found in our blood. Moreover, the stock of progenitor cells needs to be maintained in the appropriate conditions for further use in the differentiation of billions of blood cells, and for this, maintaining the appropriate microenvironment is essential. The microenvironment of a bone marrow stem cell is called a 'niche' and is itself constituted by the local tissue microenvironments responsible for maintaining and regulating the stem cells. They are characterized by different physical properties, proteins and cell types: hematopoietic stem cells*, osteoblasts*, endothelial cells* and bone marrow stromal cells* (macrophages, adipocytes fibroblasts). This environment plays an important role in cell activation and differentiation. Bone marrow is composed of three of these regions or niches: the endosteal, the central and the perivascular niche.

Bone marrow has been an exemplary model for the struggle to imitate tissue function in three-dimensional cell cultures. The impossibility of modeling tissue interaction has in the past always represented a seemingly insurmountable obstacle for understanding the molecular mechanisms of disease in this context. Researchers have been able to identify and describe a large variety of microenvironments but only the challenge of constituting a functional bone marrow on a chip has been sufficient motivation for addressing the question of what is essential to these environments and in what circumstances they can be seen to fulfill which functions.

The long-term culture of progenitor cells has been carried out in three-dimensional collagen* scaffolds perfused with nutrients, but the three different niches that constitute the physiological context for bone marrow have not yet been brought together on a single chip. In addition, varying oxygen pressure is difficult to master in these circumstances. Thus, this kind of system demonstrates how difficult it is to reduce the complexity of an organ where one needs to control the nutrients available, integrate different matrices for modeling different niches and so come up with a highly complex integrated device. The problems confronted by attempts to put bone marrow on a chip illustrate well how oversimplification can compromise the functioning of integrated physiological systems. Still, the positive outcomes of these attempts illustrate the interest of persisting in this research, even though the organ-on-a-chip has not yet reached its full potential. This research has allowed scientists to mimic various leukemia* (a bone marrow pathology affecting the white blood cells) cells for testing the efficacy of new drugs or measuring sensitivity to radiation. These models have also opened up the possibility of studying angiogenesis and cell adhesion. Finally, albeit on a rudimentary level, these leukemia models have allowed studies of the metabolism of cytostatic* drugs and chemotherapy resistance, opening up new opportunities for being able to study bone marrow disease. So, while a satisfactory model of bone marrow on a chip has not yet resulted from this research, other results underline its utility and encourage more projects in the area. With the principle in place, researchers are on the lookout for mechanisms that will bring these model organs closer to the real thing. One of the leading groups in this domain is based at the Wyss Institute at Harvard University, where Donald Ingber has already founded a start-up company *Emulate* to produce and market a range of such systems.[1]

The target markets for these organs on chips are pharmaceutical companies or their sub-contractors specialized in the research for new leads, that is to say looking for substances that appear to be promising for medical use. These companies test hundreds of thousands if not millions of molecules from their extensive chemical 'libraries' to try and identify a handful that appear to be effective against target diseases and can be further developed as potential medical drugs. While the preliminary tests, or screens, are generally already carried out on isolated cells rather than animals, follow-up tests use large numbers of animals, and it is here that these organs on chips are meant to come into their own. They are intended to replace animals at these earlier stages of drug development, and, ideally, all the subsequent tests right up to the moment where human subjects can be used. In light of the number of animals that are used for experimentation every year, it is a tall order to replace them all with appropriate artificial organs on chips.

7.9 ORGANS ON CHIPS AND THE NANO

Organs on chips represent a reductionist modular conception of biological complexity. A change of paradigm, as a whole animal or any part of it that can function as a system—pulmonary, hepatic, nervous or vascular—can be modeled by in vitro systems. More or less complex physiological processes can be integrated and function together in a small device made up of modular subunits. This implies a reduction in scale comparable to the reduction from micrometric to nanometric size. Structures and microfluidics* that are the functional elements of these artificial organs are built from elastomers* using photolithography* with a resolution comprised between 20 and 100 nm. But, while these elements clearly fall under the definition of the nanosciences, are the organs on chips themselves nano?

As with the microarrays we were looking at in a previous chapter, these chips are not themselves strictly speaking nano in scale, but they mobilize a number of techniques from the nanosciences and are often, as we said, considered to be part of nanomedicine. Unlike microarrays, however, the biological complexity and integrity of these model organs make it unlikely if not impossible that the objects themselves will

[1] More information is available on their website — https://wyss.harvard.edu/ (accessed 15 June 2022).

ever be reduced to an authentically nano scale. This is even more the case when cells from different organs are connected together via microfluidic channels to form a more sophisticated integrated model combining organs into systems and even potentially providing a 'human on a chip', which would usher in a profound change in drug testing in general and precision medicine in particular. Are we, therefore, cheating a little in our book on nanomedicine by remaining at the level of micro-reduction when presenting this promising new technology of organs on chips? In fact, we want to exploit this ambiguity between the authentically nano and organic models that are highly reduced in size. The example of this micro-technique of organs on chips provides us with the chance to approach nanomedicine from a different angle, thinking in terms of the goals and mechanisms of 'miniaturization' rather than starting directly at the nano level to qualify an element of nanomedicine.

7.10 WHY DO WE NEED ORGANS ON CHIPS?

Animal experimentation, first in the context of physiological research and later in the pharmaceutical industry, has always been the target of popular protests. A number of groups continue to campaign against the testing of pharmaceutical products on animals. While in Europe the practice of animal testing has already been outlawed for cosmetics, it continues in the wider pharmaceutical industry, particularly in the context of drug discovery and development. Despite a high level of public disapproval,[2] animal testing continues because of a lack of reliable alternatives across the board for preclinical tests, although the problem is most pressing in neurological and behavioral studies, where organs or even organisms on a chip promise very little.

The context of a turn against animal experimentation has led to legislative changes in Europe in particular, and the inscription into the law in 2010 of the three 'Rs': Replacement, Reduction and Refinement. This approach requires researchers to try and replace animals in their scientific experiments wherever possible and when they are unable to do so, to reduce the number of animals that are used.[3] Refinement refers to minimizing animal suffering by working on the experiments that are done so that the animals do not suffer. While it started life in the 1960s as an academic initiative, the 3 Rs movement enjoys broad-based support.

In 2010, the European parliament introduced a directive aimed if not at eliminating animal use in scientific research, then at least reducing their number. The directive states that 'The use of animals for scientific or educational purposes should therefore only be considered where a non-animal alternative is unavailable'.[4]

Thus, there are legislative pressures for limiting animal testing, providing a notable incentive for the development of organs on chips. This being said, there is every reason to believe that research scientists would like to reduce if not eliminate the use of animals in this research as well. If it is unnecessary, why continue testing

---

[2] Institut d'études opinion et marketing en France et à l'international (IFOP) estimated opinion as being 90% opposed to all animal testing in 2018.

[3] www.nc3rs.org.uk/the-3rs.

[4] Directive 2010/63/EU of The European Parliament and of The Council, 22 September 2010—on the protection of animals used for scientific purposes.

new drugs on animals? Who would advocate animal suffering for its own sake? Or maybe, as certain critics have suggested, the disaffection with animal testing among scientists comes from elsewhere. Two such reasons have been suggested: the number of false leads in drug development, and the difference between human and animal models in terms of both drug action and toxicology. The problem of false positives is inherent in any experimental system that relies on statistical reasoning. With machines capable of testing millions of molecules every year against target diseases, statistically, the algorithms used to process the results will indicate a certain number of these chemicals to be active when they are not; each one of these results is a false positive. Of course, this phenomenon is well known, and there has been considerable investment to develop various in silico approaches that use computer modeling to identify the most promising molecules. Three-dimensional digital modeling and representation of both the target molecules (receptors, enzymes, etc.) and the therapeutic drugs of interest are also constantly evolving, allowing researchers to perform quantitative structure–activity relationship (QSAR) analyses that help in identifying promising directions for new leads independently of animal testing (Peter et al., 2019). Thus, these QSAR techniques should also reduce the number of tests that are performed on animals, serving to orient choices within any drug screening program.

While working on the screening methods can potentially reduce (although not eliminate) any such false positives, the problem of the physiological differences between humans and the animals used in pharmacological tests is harder to resolve. Neither a rabbit nor a mouse is a human being, and despite attempts to genetically engineer or specially breed animals to better model human physiology, the differences between the animal model and a human are the source of a number of false leads. Here, we are not in the domain of simple statistical false positives, but in the area of irreducible differences between humans and other animals. Toxicological studies on the etiology of cancer have well characterized the difference in carcinogenic potential between the different metabolisms of animals. Of course, for ethical reasons, it is inconceivable to test promising drug leads on human subjects without prior animal testing, but could this animal testing be replaced by a system using organs on chips? Particularly if these organs on chips give a better indication of the efficacy of a drug against a targeted disease in humans.

We also have to consider that the impetus behind the reduction or elimination of animals from contemporary testing models might not be straightforwardly ethical, but rather economic and maybe even primarily economic; in the end, using animals is just too expensive, and this is due to the same paradox we saw in Chapter 6. Even the much less costly use of cells to try and identify potential leads for new drugs is proving expensive due to the high throughput methods that multiply the number of tests being done. We have already seen with our discussion of microarrays how miniaturization and eventually reduction to the nanoscale have contributed to the multiplication of such tests, and there is no reason to expect that this trend will change. Of course, in this context, the rapid expansion of computing power has been another major impetus behind the increase in the number of tests. Thus, the organ-on-a-chip has attracted the attention of pharmaceutical companies as a way to economize on this form of research. So, we can see that this development is potentially beneficial for pharmaceutical research on several levels at once: saving animal lives, saving money, saving on materials and, the key element, giving a more realistic assessment of which drugs under test might work in human beings. We can, of course, see once again the alignment with the precision medicine approach, as the organs on chips can be constituted from the cells of a specific patient, and the reaction to the various treatments under consideration (both positive and negative) can be tested directly on this miniaturized model of the patient's own body, giving doctors further information to help refine the appropriate choice of treatment.

Today, organs on chips constitute a highly promising research field that will allow researchers to avoid the use of many if not all of the laboratory animals, particularly rodents, in the near future. Since 2010, a number of these devices have already been proposed by companies in ready-to-use packages like lung-, nerve-, heart-, liver- and even the body-on-a-chip that combines up to nine organs.

We have already seen the high-volume micro-array systems that have saved thousands if not millions of animal subjects from testing. Indeed, a whole range of *in vitro* and *in silico* tests reduces the need to engage these animals in experiments to determine potential drug activity as well as harmful toxicity. This being said, the multiplication of research projects and the broadening of the bases for searching for new leads could well mean that overall just as many animals are subjected to these kinds of tests. This is yet another reason why it is so important to find techniques that can replace the use of these animals entirely.

One of the approaches that has been actively promoted is the computer modeling of toxicology and drug action. And computer simulations are clearly important indicators for orienting the choice of prospective leads for developing new drugs. Of course, *in silico* testing raises its own concerns in terms of the quality of the representation of complex living organisms by their virtual avatars. But this question of representativity will always be present when researchers adopt any kind of substitute for a test on a living organism, and, as we have already pointed out, even animal models pose the question of representativity for modeling physiological action in humans.

7.11 TUMORS ON CHIPS

A recent but related development in the same area as organs on chips is the tumor-on-a-chip, available in either a flat, two-dimensional format or a three-dimensional conformation. This kind of chip mobilizes a miniaturized space to perform (i) physiopathological and biological mechanistic studies on disease, (ii) drug screening of multiple tumors under different physiological conditions and (iii) the implementation of personalized or precision medicine using lab-on-chip studies. Here are two examples drawn from oncology, one concerning breast tumors and the other, brain metastases.

The brain metastasis of a cancer generally comes with a fatal prognosis, associated with a median survival time of one year. Even if the original tumor is well-identified, the blood-brain barrier makes it difficult for chemotherapy to reach metastases in the brain, rendering chemotherapy ineffective. Brain niches can be mimicked by sequentially seeding astrocytes and endothelial cells that play the roles of brain tissue and the blood-brain barrier, respectively. On the glass slide of a microscope 16 samples of tumor cells (e.g., breast cancer cells that migrate frequently to the brain) can run simultaneously on such a three-dimensional microfluidic brain niche. Confocal microscopy can determine the migration distance of cancerous cells in neural tissue (astrocytes) from the endothelial plane over a few days. The greater the distance traveled, the greater is the migration capability of the cancerous cells: the migration of cancer cells across a brain barrier is called extravasation, one of the parameters for determining the metastatic potential of a tumor. It is also possible to measure the total volume of migrated cancerous cells, giving a further indication of the capacity of a tumor to induce brain metastasis. Of course, such studies have been performed in the past using 3D models provided by the Boyden chamber*, but these new studies are repeatable, more accurate and easier to perform using tumor on chip technology and confocal microscopy. By coupling both techniques with a machine learning program resulting from research in artificial intelligence, this preclinical model of a tumor environment can be automated. What in the past was the prerogative of a specialized research laboratory is now feasible in a generalist hospital biology laboratory.

The possibilities associated with the development of these techniques are endless—as are the potential tie-ins with precision medicine. One example is the use of breast cancer cells, such as ductal carcinoma in situ. Here, cells can be obtained from a patient using a biopsy and put on a layer of mammary cells mimicking a mammary duct layered onto stroma composed of collagen and fibroblasts, both components of a synthetic extracellular porous matrix that separates two chambers in which anti-cancer drugs can be introduced to study the susceptibility of a given tumor to drug action. Another use of these chips is to test samples for reactivity to hormones and growth factors. Moreover, some breast-on-chip platforms allow researchers to investigate the role played by crucial parameters associated with the environment—hypoxia* being one example—on the intravasation process of breast cancer cells. Endothelial* cells surrounded by fibroblasts* can be used to construct capillary vessels and microfluidic devices allow their perfusion with breast cancer cells. Counting the cells migrating in the vasculature allows researchers to attribute an extravasation rate to each tumor, thus enabling them to estimate its metastatic potential, especially in conditions of hypoxia.

Industrial actors are very interested in this group of techniques, and the global microfluidic market continues to grow. It is expected to reach US$28 billion in 2023 in part due to a growing chip segment. Just to give one example, the Dutch company Mimetas™ sells an OrganPlate® platform that allows the culture of 96 micro-tissues for patient-derived cancer cells or tissue screening. Tumors on chips are also useful for disease and especially cancer stratification, an objective established some 20 years ago with the development of both omics and transcriptomics*. It is not inconceivable that the DNA of a solid tumor can be entirely sequenced and checked for genetic markers for growth, migration and metastasis and drug susceptibility. Once isolated in microfluidic systems, the circulating tumor cells (CTCs) could be tested and compared with the parental tumor from which they derived in order to analyze the genetic variation of these circulating cells. The aim of all this research is to obtain more effective therapies against both solid and 'liquid' tumors.

7.12 MINIATURIZATION AND THE NANO

The organ-on-a-chip is all about miniaturizing, with the ambition of supplying a functioning model of a group of interconnected organs or a whole human body on a series of interconnected chips if not on one single organ system on a chip. This modular or integrated construction of a model human would allow the introduction of a cancerous tumor or whatever else might be under investigation into a specific organ but no longer in isolation, as the chip or dedicated zone of a chip would now be integrated into the whole model body. Furthermore, when the cells can be drawn from the prospective patient herself or himself, the results of tests will be even more pertinent (once again making a link with the precision medicine movement).

Nevertheless, these simplified models of organs and multi-organ systems although much smaller than a living animal will never arrive at the nano-scale for reasons we discussed earlier. And yet, as we have already mentioned, the organ-on-a-chip is often included as an illustrative example in overviews of nanomedicine. What are we to make of this paradox—if not confusion—that seems to conflate the very process of miniaturization with the nanosciences? In previous chapters, we considered nano-pharmacy and discussed nanodiagnostics, which mobilize nanoparticles for medical goals. While not yet, and for the reasons we have already set out here probably never formally a nanotechnology, the 'organ-on-a-chip' is the product of the same logic of miniaturization. Given the continuities of approach and the artificiality of the strict 100 nm border around the nano, it is useful to consider these different approaches like microarrays and organs on a chip together with the use of probes and nanovectors that answer to this stricter definition.

7.13 NANOMEDICINE AS A DISTRIBUTED APPROACH TO MEDICAL SPECIALTIES

Following up on this question of miniaturization and the definition of the nano, we can also pose the question of the status of nanomedicine as a specific domain of research or even a distinct scientific discipline. In all the examples we have looked at, from nanopharmacy to organs on chips, the cross-disciplinary application of a nano or other miniaturization approaches has quickly narrowed down to the investigation of an application in one single medical domain. Often, the examples have been drawn from oncology, the medical specialty concerned with cancers, but we have also considered organ systems that would take us into the area of nephrology (the kidneys), neurology (the central nervous system) or hematology (blood). The point is that the use of nanoparticles or other nano-techniques is likely to apply across the range of medical care. For the moment, nanoinitiatives tend to attract financing for cross-disciplinary projects, and yet the applications that develop out of innovations in the domain of the nano are local and often confined to individual medical specialties.

We do not have the space here to decline all the specialties that have developed some connection with the nano: nanooncology, nanonephrology, nanohematology, etc., nor do we really need to. Instead, we should reflect on how this transversal approach could well end up impinging on most, if not all medical specialties. In our previous chapters, particularly the one on nanotheranostics, we gave considerable space to examples in oncology, but this is far from being the only medical specialty that has engaged with nanoscience and nanotechnology, or even with nanotheranostics. There are, of course, good reasons why cancer is so prominent in these applications of the nano, which, we would argue, tells us more about cancer research than it does about the nanosciences.[5] But we also need to take a broader look across the board in medicine, because there will be few areas of medicine that will not be touched in some way by the nano.

Today's medical specialties, and there are a large number of them, are the result of a trend that began in medicine in the nineteenth century, particularly in Europe, and has yet to end (Weisz, 2006). Childbirth, imaging, individual organs and classes of disease like cancer or infectious diseases have all served as the basis for the constitution of such specialties, and whenever a particular diagnostic or therapeutic approach attains a critical mass, it stands a good chance of generating another one. Nevertheless, given the nature of nanomedicine, it seems unlikely that it will produce another independent medical specialty—nanomedicine—where practitioners specialized in the field are recruited into a single department of a university or hospital and belong to the same professional bodies. While it is true that there are now several journals dedicated to nanomedicine, another sign of an emerging discipline, the majority of articles on the subject are published in the specialist journals of other areas of medicine.

The nano is not, then, specialty based, and yet it will affect most areas of medicine; it is perhaps better to conceive of it as a resource or set of resources that will be mobilized across a range if not all medical specialties. This means that the nano are liable to transform medicine across the board in a similar fashion to the techniques used for medical imagery. Magnetic resonance imaging (MRI) and positron emission tomography (PET), for example, are techniques drawn from physics that have clearly changed the practice and function of medical imagery in the second half of the twentieth century, and thanks to the increasing diagnostic power of these techniques, they have left no specialist area untouched.

Medical specialties can be understood a number of different ways. Today, the most common approach would be to see such a specialty as a 'logical' if not 'inevitable' outcome of the growth of knowledge in biology, medicine, and all the accessory sciences, notably genetics. With the large number of publications in every domain of medicine, it is difficult even for the specialists to stay up-to-date with research in their own area, making it unthinkable to try to master two different specialties.

Apart from oncology, we have not really considered medical specialties, and yet these specialties not only structure modern medical treatment, but they also inform much of the scientific research in medicine. So what kinds of innovations are the nanosciences liable to contribute to each specialty? By definition, each specialty is different and so, as with the imaging techniques mentioned earlier, the nature of the contribution of the nano is liable to vary. We have already seen some cross-specialty approaches—diagnosis, arrays, probes. And the applications

---

[5] Because of the poor prognosis associated with many cancers, they tend to attract clinical trials testing innovative approaches. It is a domain where many, particularly those suffering from advanced or metastatic cancers, are willing to participate in clinical trials of new drugs or other approaches, as they offer a chance of living longer than with palliative or other established care protocols.

of the nano to cancer chemotherapy are relevant to other specialists who regularly confront cancers—hematology, gynecology, gastroenterology, and reproductive medicine, just to mention a few. But each specialty will have its own niche nanomedicine as well. An article on reproductive medicine from 2014 points not only to the theranostic approaches in prostate cancer, but also in endometriosis, uterine fibroids, and ectopic pregnancy. It is hoped that the nano will be able to contribute significantly to both assisted reproduction techniques (ART) and in vitro fertilization (IVF). Beyond these techniques already in use in the treatment of human infertility, there are procedures that are being developed for animal breeding, such as sperm-mediated gene transfer and cell sorting, techniques that could potentially be applied to help prepare genetically compatible xenotransplants, from pigs to humans, for example. More realistically, at least in the short-term, the authors underline the importance of the nano in their contribution to the biological detection of disease conditions, medical imaging, drug delivery and tissue engineering (Barkalina et al., 2014).

We can imagine an equivalent analysis of any other medical specialty and, while it is true that we have tended to focus on cancer—almost always the first sub-domain to be mentioned in reviews in other specialties, like the one on reproductive medicine—it is hard to imagine that any area of medicine will not be affected by the developments we have discussed in the previous chapters.

The excitement around nanomedicine at the end of the twentieth century and at the beginning of the twenty-first century has now receded—but this abatement if not decline in interest in 'nanomedicine' should not lead us to the conclusion that the techniques we have already discussed will disappear or that new ones—which could qualify as nanomedicine—will not be developed. Of course, some will never be developed at all, and others will branch off in unexpected ways, but globally, the movement we have been presenting as nanomedicine in this book is here to stay. What we have been suggesting in this chapter is that increasingly these developments, which could be identified as nanomedicine, will be integrated into existing medical specialties like dentistry or infectiology depending on its area of application. Thus, our speculative conclusion is that there is no guarantee that any of the techniques or approaches discussed so far will continue to be identified as nanomedicine. Instead, nanomedicine risks being distributed throughout all the medical specialties as a 'nanobranch' of the domain in question.

7.14 TO GO FURTHER

As references to go further in the domain of organs on chips, we propose a general discussion of miniaturization (Peercy, 2000), and overviews of the organ-on-a-chip and its use in the evaluation of toxicity (Azizipour et al., 2020; Pun et al., 2021; Wu et al., 2020; Kratz et al., 2019; Cong et al., 2020). Here are some leads to go further into specific organ-on-a-chip models such as bone marrow (Santos Rosalem et al., 2020), breast cancer (Song et al., 2021; Subia et al., 2021) and skin (Graff et al., 2022).

Azizipour, N., Avazpour, R., Rosenzweig, D. H., Sawan, M., & Ajji, A. (2020). Evolution of biochip technology: A review from lab-on-a-chip to organ-on-a-chip. *Micromachines, 11*(6), E599. https://doi.org/10.3390/mi11060599

Cong, Y., Han, X., Wang, Y., Chen, Z., Lu, Y., Liu, T., Wu, Z., Jin, Y., Luo, Y., & Zhang, X. (2020). Drug toxicity evaluation based on organ-on-a-chip technology: A review. *Micromachines, 11*(4), E381. https://doi.org/10.3390/mi11040381

Graff, P., Hönzke, S., Joshi, A. A., Yealland, G., Fleige, E., Unbehauen, M., Schäfer-Korting, M., Hocke, A., Haag, R., & Hedtrich, S. (2022). Preclinical testing of dendritic core-multishell nanoparticles in inflammatory skin equivalents. *Molecular Pharmaceutics, 19*(6), 1795–1802. https://doi.org/10.1021/acs.molpharmaceut.1c00734

Kratz, S. R. A., Höll, G., Schuller, P., Ertl, P., & Rothbauer, M. (2019). Latest trends in biosensing for microphysiological organs-on-a-chip and body-on-a-chip systems. *Biosensors, 9*(3), E110. https://doi.org/10.3390/bios9030110

Peercy, P. S. (2000). The drive to miniaturization. *Nature, 406*(6799), 1023–1026. https://doi.org/10.1038/35023223

Pun, S., Haney, L. C., & Barrile, R. (2021). Modelling human physiology on-chip: Historical perspectives and future directions. *Micromachines, 12*(10), 1250. https://doi.org/10.3390/mi12101250

Santos Rosalem, G., Gonzáles Torres, L. A., de Las Casas, E. B., Mathias, F. A. S., Ruiz, J. C., & Carvalho, M. G. R. (2020). Microfluidics and organ-on-a-chip technologies: A systematic review of the methods used to mimic bone marrow. *PloS One, 15*(12), e0243840. https://doi.org/10.1371/journal.pone.0243840

Song, K., Zu, X., Du, Z., Hu, Z., Wang, J., & Li, J. (2021). Diversity models and applications of 3D breast tumor-on-a-chip. *Micromachines, 12*(7), 814. https://doi.org/10.3390/mi12070814

Subia, B., Dahiya, U. R., Mishra, S., Ayache, J., Casquillas, G. V., Caballero, D., Reis, R. L., & Kundu, S. C. (2021). Breast tumor-on-chip models: From disease modeling to personalized drug screening. *Journal of Controlled Release: Official Journal of the Controlled Release Society, 331*, 103–120. https://doi.org/10.1016/j.jconrel.2020.12.057

Wu, Q., Liu, J., Wang, X., Feng, L., Wu, J., Zhu, X., Wen, W., & Gong, X. (2020). Organ-on-a-chip: Recent breakthroughs and future prospects. *Biomedical Engineering Online, 19*(1), 9. https://doi.org/10.1186/s12938-020-0752-0

7.15 REFERENCES

Barkalina, N., Charalambous, C., Jones, C., & Coward, K. (2014). Nanotechnology in reproductive medicine: Emerging applications of nanomaterials. *Nanomedicine: Nanotechnology, Biology, and Medicine, 10*(5), 921–938. https://doi.org/10.1016/j.nano.2014.01.001

Peter, S. C., Dhanjal, J. K., Malik, V., Radhakrishnan, N., Jayakanthan, M., & Sundar, D. (2019). Quantitative structure-activity relationship (QSAR): Modeling approaches to biological applications. In S. Ranganathan, M. Gribskov, K. Nakai, & C. Schönbach (Eds.), *Encyclopedia of bioinformatics and computational biology* (pp. 661–676). Academic Press. https://doi.org/10.1016/B978-0-12-809633-8.20197-0

Risueño, I., Valencia, L., Jorcano, J. L., & Velasco, D. (2021). Skin-on-a-chip models: General overview and future perspectives. *APL Bioengineering, 5*(3), 030901. https://doi.org/10.1063/5.0046376

Theobald, J., Ghanem, A., Wallisch, P., Banaeiyan, A. A., Andrade-Navarro, M. A., Taškova, K., Haltmeier, M., Kurtz, A., Becker, H., Reuter, S., Mrowka, R., Cheng, X., & Wölfl, S. (2018). Liver-kidney-on-chip to study toxicity of drug metabolites. *ACS Biomaterials Science & Engineering, 4*(1), 78–89. https://doi.org/10.1021/acsbiomaterials.7b00417

Weisz, G. (2006). *Divide and conquer: A comparative history of medical specialization.* Oxford University Press.

8 Regenerative Medicine
Mobilizing the Body's Own Repair Mechanisms

Regenerative medicine is another emerging field in contemporary healthcare that has integrated the nanosciences, or more specifically that has incorporated a range of new nanomaterials into its arsenal. The principle of regenerative medicine is quite straightforward, as it aims at the repair or replacement of damaged tissue or even organs by employing the activity of living cells. Described like this, regenerative medicine might seem quite banal; after all, much of modern medicine already pursues the same goals by different means. One only needs to think about transplant surgery that has saved millions of lives by replacing vital organs—hearts, lungs, livers and kidneys—with ones received from donors, or the use of prosthetics to replace failing joints and amputated limbs, to see that many medical specialties already aim to replace or repair damage caused by wounds, accidents, wear associated with aging, or disease. The key difference in regenerative medicine is the mobilization of the body's own cells, and in particular stem cells (pluripotent, multipotent or even, in principle, totipotent cells—we will explain the difference in what follows) as the active basis for the repair process. Still the parallel between transplant and regenerative medicine is helpful, as it gives us a clue as to why the latter has provoked so much interest on the part of philosophers.

Reparative surgery is a form of plastic surgery, but when people think of this latter term, it is generally cosmetic plastic surgery that comes to mind: facelifts, liposuction, breast implants or hair transplants. The two forms of surgery employ the same techniques and are often performed by the same surgeons, meaning that we might well have to look to the motivation for a given surgical intervention in order to classify it as either reparative or cosmetic. In the case of cosmetic surgery, patients are looking for surgical assistance to change (usually to improve) their appearance, although whether this stems from a superficial desire for change or from a deep-seated psychological motivation can make a significant difference to how people from the exterior perceive and judge these interventions. This range of plastic surgery—from repair to enhancement—already suggests what direction regenerative medicine could potentially take in the future, and this prospective development explains in turn why it has excited interest outside the specialist sphere of its medical applications. Regenerative techniques that go further than well-defined therapy can potentially go a very long way in this direction, leading beyond just repairing body parts to their augmentation. Medical teams will be able to endow organs with improved function and increased capacities, no longer just returning patients to minimal functionality, or to a previous baseline standard, but adding to what they were before. Regenerative medicine has

DOI: 10.1201/9781003367833-8

this potential for enhancing the body, and maybe the mind as well. We announced in our introduction that we would do our best to avoid the issue of transhumanism, even though it is a favorite topic for philosophical discussions about the potential impact of nanotechnology on our world. The image of nanomachines or nanorobots represented as little mechanical humanoids circulating in our bloodstream, fixing or replacing any deficient or underperforming cells has been integrated into the popular imagination of the nanosciences, no matter how unrealistic it might be. But even though the mechanism is a little far-fetched, the principle of nanotechnology intervening invisibly to repair and renew our bodies is not so fantastic. The idea that piques the philosopher's interest is that intervention at the nano level will not only be able to prolong healthy life and improve human capacities (both physical and, more significantly for philosophers, intellectual) but also ultimately transcend the human condition, in principle enabling individuals to become immortal. As death is the only certainty that we all share in common, ideas—or should we term them fantasies—about escaping this universal fate have permeated human culture throughout its history. Immortality is generally what is taken to separate gods from humans, and a heightened faith in technology inspired by the considerable changes it has already brought about in the modern world has led a small but vocal group of thinkers to imagine that modern science and technology will soon be able to work this transformation. Transcending the human condition is an ambition that lies at the heart of the transhumanist philosophy, and the nanosciences have recently been identified as a leading candidate for contributing to, if not actually bringing about this transcendence. Thus, when discussing regenerative medicine, it seems impossible to escape at least some of the questions raised by the transhumanist movement, although we shall steer clear of the most radical speculations about human immortality.

8.1 THE POTENTIAL OF THE STEM CELLS BEHIND REGENERATIVE MEDICINE

Regenerative medicine as a novel approach to repairing damaged tissue has its roots in the twentieth century, and the discovery and mobilization of stem cells as a way to produce differentiated animal tissues. Stem cells are, along with our DNA, a major part of the answer to one of the greatest mysteries of life: how do fully grown adults with differentiated cells composing all their different body parts come out of a fertilized egg—one single undifferentiated cell? Stem cells are similarly undifferentiated cells on the path of the development of an organism that have the potential to form different types of tissue and, following genetic and environmental cues, give rise to differentiated cells of a particular type. These stem cells are classed into pluripotent, multipotent and totipotent cells depending on the range of the types of cells that they can produce. Totipotent stem cells have the potential to produce any kind of cell in the body (of course, this is the case of the fertilized egg), pluripotent cells can become almost any cell in the body, but are limited to those beyond the embryo stage, while multipotent stem cells are the most limited category of stem cells, having the potential to become different types of cell, but within a given tissue, such as blood, muscle or, what will interest us in the example we will be considering here, bone. The potential of this approach becomes clear if we consider the case of the fertilized egg of a human being: a single cell that by differentiated reproduction is capable of producing over 200 types of tissue in an adult body. If this potential could be tapped in other contexts, and it is not inconceivable that other cells placed in the right conditions can become totipotent themselves, then reproductive medicine as well as regenerative medicine would considerably expand their horizons. Techniques for inducing multipotency in ordinary cells have constituted a very active area of research in stem cell science, but we will have more to say about the potential of these developments when we return to stem cell research later in this chapter. For the moment, it is enough to note that over the course of the twentieth and into the twenty-first centuries, stem cell research has made considerable headway, allowing researchers to conceive of more sophisticated techniques in regenerative medicine. Nevertheless, it is only recently that the nanosciences and nanotechnology have started making significant contributions to this area.

As with much of the sprawling domain of nanomedicine we have been considering up until now, the use of the nanosciences and nanotechnologies (the nano) in this area has, up until now, largely been confined to research laboratories. But, once again, the potential of the application of the nano to this field of regenerative medicine is as large as the range of nanomaterials already available today and potentially available tomorrow.

So, regenerative medicine covers the range of medical techniques that aim to compensate for the loss of function in organs, the loss or dysfunction of tissue or even the loss of entire organs, by deploying living cells, preferably originating from the patient herself or himself, in order to replace the living material that has been lost. Thus, its development has largely depended on that of stem cell research, seeking to channel and enhance the multipotent qualities of activated or re-activated stem cells toward growing specific tissue, parts of organs or even whole organs either regrown in situ or transplanted from a surrogate host. Before giving a fuller account of this stem cell research, we would like to consider the more mechanical or architectural contribution of the nanosciences to regenerative medicine. There is a growing research community that is examining the potential of using nanomaterials to structure regrowth in damaged areas of the body. Taking bone as our exemplary organ, we will look at what is being proposed in the area of nano-based bone replacement today. We will then return to the significance of linking the nanosciences to stem cell research in a second part of the chapter, and so consider some other areas of regenerative medicine that are being explored. As we shall see, the use of nanomaterials combined with stem cell technology opens the door to certain aspects of the transhumanist agenda, particularly around the issue of human enhancement, so we will reserve some space for these considerations at the end. We will not go very far along the path of these transhumanist reflections, because once you have started down this speculative path, it is very hard to know where it might lead.

8.2 REPLACING MISSING BONE AS A NANO-ARCHITECTURAL, FUNCTIONAL CHALLENGE

Before considering a wider range of regenerative medicine, we will explore the example of replacing damaged or missing bone, one of several active areas of research in the field, in more detail. When bone needs to be replaced, it is still largely done using bulky metallic prostheses; this is particularly the case for joint replacement, with hundreds of thousands of implants based mainly on titanium processed alloys and ceramics replacing hips and knees every year all over the world. Nevertheless, putting these joint replacements to one side, the favored method for replacing or repairing damaged bone is by grafting it from elsewhere in the patient's body. The use of a so-called autograft* has the advantage of not introducing 'foreign' matter (from a donor) into the body, and so reduces the drawbacks of having to prepare the materials from a donor, stripping them as much as possible of their antigenic features, in order to make a successful transplant. The autograft is not, however, without drawbacks or risks, as harvesting the bone from elsewhere in the body weakens the donor site and can lead to long-term health problems. Thus, other materials, notably synthetic polymers and ceramics are also used to help repair damaged bone, but the problem here is that these artificial materials lack certain characteristics of the natural bone that they are being asked to replace. After all, bone is a remarkable structural material, striking a delicate balance between rigidity and flexibility, with the skeleton providing a rigid structure around which the animal body is constructed, while its constituent bones remain flexible enough to resist the impacts and shocks associated with a healthy, active lifestyle.

Many believe that regenerative medicine holds the key to such functional repair, mobilizing the body's natural repair mechanisms extended beyond the limited context of this natural healing. The appropriate deployment of nanomaterials in this context will, it is hoped, provide a kind of intermediate solution between synthetic prosthetic approaches and the natural healing process, leading to an effective but less invasive method for repairing bone or just accelerating this natural repair process. In this context, nanomaterials provide structural support for new bone growth, but more than this, these materials can be functionalized by using known sequences of peptides, among other techniques, to stimulate the cells that adhere to them, supporting and reinforcing the natural regenerative functions of the stem cells that lie behind the new growth.

As we just remarked, the basis of regenerative medicine is the imitation or rather the extension of natural tissue repair processes, which means that developing appropriate strategies has relied on prior research into the structure and functioning of the living tissue that is going to be replaced or repaired. This kind of fundamental anatomical research into the details of bone physiology itself reveals a complex integrated living material. Bone is structured by an extracellular matrix* made up of a range of larger molecules, the most important (in terms of their presence) being hydroxyapatite and collagen. This matrix not only gives the bone its form but also provides the environment in which the constitutive bone cells can live and reproduce. Bone remains healthy and functional in an adult body thanks to a process of homeostasis, where the materials lost from the bone (through lysis favored by osteoclasts) are replaced in permanence (a regenerative process driven by the osteoblasts that evolve into osteocytes and are capable of synthesizing the hydroxyapatite crystals as well as the rest of the bone matrix), maintaining a constant composition and ensuring that the bone retains its structural integrity, preserving its characteristic balance between rigidity and flexibility.

It is the structuring role of this matrix that has inspired medical researchers to search among the growing range of synthetic nanomaterials in order to find one (or a combination of several) appropriate for forming an artificial matrix on which the new bone cells can grow, and so replace the bone grafts we discussed earlier. The term used in regenerative medicine for the structure in which the bone is going to grow is the scaffold. It is not sufficient, however, for the scaffold simply to play the structural role that its name suggests; it also needs to be attractive (or at least hospitable) for the cells that will be at the origin of the replacement growth of the cellular material and to channel the rest of the renewed bone into an appropriate conformation in line with the goals of the tissue repair. Various techniques are used to shape the implanted scaffold, generally grouped under the head of rapid prototyping techniques. Furthermore, the surfaces of the material making up the scaffold should at the very least enable if not encourage the bone regeneration process. A number of growth-stimulating factors are used in order to favor the growth of new bone, introducing another element into the equation, but more of this in what follows.

In the domain of regenerative medicine for bone, several different nanomaterials are now being tested as candidates for this kind of scaffold. After the new tissue has formed, the scaffold, as the building metaphor suggests, will be taken away, but even though it is destined to disappear, it nevertheless makes an important contribution to the process of reconstituting the damaged area in two ways. It provides both structural form for the growth to take place and appropriate surfaces for encouraging the installation and multiplication of the cells that will be doing the regenerative work. So which nanomaterials are currently seen as candidates for preparing an appropriate scaffold for regenerating bone?

8.2.1 Nanoparticles, Nanotubes and Nanofibers

It is perhaps surprising, and another sign of the diversity of approaches that make up nanomedicine that it has taken us this long in a book on the subject, to talk about the contribution of the nanomaterials that are already well-established industrial products—carbon nanotubes and graphene—except as potential environmental hazards. These nanoparticles, which are contributing to the contemporary material culture of the nano in plenty of other domains, are among the many candidates being considered for the scaffolds that could guide and facilitate bone replacement. Although it is far from clear what role these two nanomaterials will ultimately play in bone replacement therapy or in regenerative medicine more widely, we will take this opportunity to look at them in more detail as part of a presentation of the range of materials under investigation for scaffold in the regenerative treatment of bone loss or injury. This being said, in the spirit of biomimicry that has inspired much of regenerative medicine, it is likely that composites integrating particles of hydroxyapatite—as we have seen, one of the principal constituents of bone—will be favored as materials for this research.

8.2.2 Carbon Nanomaterials

One of the remarkable and much remarked upon contributions of nanotechnology has been the multiplication of the known forms of elementary carbon, leading to the industrial production and use of pure carbon in quite novel nanostructures. The forms of pure carbon that are well known to anyone who has studied some elementary chemistry are, of course, graphite and diamond, but the nanosciences have also identified buckminsterfullerene and other spherical forms of elementary carbon (a family of molecules known as the fullerenes), as well as graphene, a planar form of carbon, and carbon nanotubes, which, as their name suggests comes in various tubular configurations. While fullerenes are used in medical applications, notably as constituents of contrast agents for medical imaging (X-ray and MRI) and have also been the subject of experimental approaches using light to stimulate their anti-cancer activity, they are not at the forefront of research in regenerative medicine. The two leading forms of carbon in this area are graphene and nanotubes.

8.2.3 Graphene—A Flagship Nano-material

Graphene is a two-dimensional form of carbon, based on flat sheets of carbon atoms that are, in principle at least, indefinitely extensible. However, graphene is not limited to these flat sheets, as the single layers can be bound together using polymers to make multi-layered versions of the material, not only giving more volume but also providing enhanced rigidity. The number of layers can be multiplied as much as is needed, and it seems logical that for providing a scaffold for bone regeneration, multi-layered graphene will be more structurally appropriate than its single-layer constituent. When put into a three-dimensional structure, the graphene can be complemented with growth factors and seeded with stem cells to stimulate regenerative growth. An excellent electrical conductor, graphene, is thereby susceptible to the influence of external magnetic fields that can also be used to stimulate regrowth in damaged areas of bone. Graphene has the additional advantage of being well characterized as a material because of its wide-ranging use in industrial products. Indeed, it has been the object of a multi-million Euro European investment plan in the form of the Graphene Flagship program[1] aimed at innovating across the board in terms of materials science. Still, graphene is just one of a number of candidates for producing scaffold for bone replacement, including its oxides.

Carbon nanotubes are another such candidate for this role, with tubes of different dimensions already being produced on a large scale for a range of industrial applications. We can find these nanoparticles in a number of everyday objects, including tennis rackets, and, as we have already mentioned, these nanotubes are the most widely used carbon-based industrial nanomaterials. These tubes are in many ways close to the graphene considered earlier, as they can be thought of as sheets of graphene wrapped around in a circle to form a hollow tube. Varying the dimensions of the tubes allows researchers to vary the properties of these materials. Historically important, these nanotubes were one of the first nanomaterials to be touted as holding the potential to transform our material world, although concerns about the toxicity of these small rods have taken away some of this luster over the years. Once again, their biocompatibility makes them viable candidates for scaffold material, with the possibility of adding growth factors and using seeding techniques similar

[1] https://graphene-flagship.eu/

to those for graphene described earlier. Of course, all these trial products need to be molded into an appropriate shape for filling the gap in the bone material, and this is often done by rapid prototyping or solid free-form fabrication, techniques that treat complex data from three-dimensional imaging in order to print out (using stereolithography, for example) a customized scaffold to fit into the damaged part of the bone with precision.

8.2.4 Peptide Amphiphiles

Another candidate for scaffold in bone is the class of peptide amphiphiles, man-made self-assembling nanoparticles first described in the 1990s. Made up of two distinct parts—one hydrophilic and the other hydrophobic—they permit self-assembly into supramolecular structures of various forms. There is considerable interest around using these kinds of materials as the basis for new approaches to this regenerative growth. Due to their chemical and physical properties, they generally fit the criteria for nanomaterials that could be used in regenerative medicine: lack of toxicity, biocompatibility and high adhesion for differentiated cells or stem cells, which, as we have already explained, are conceived as the motor for this regenerative process, but more of this in what follows. Another important consideration is the capacity of these peptide amphiphiles to undergo biodegradation without altering the host's metabolism, retaining the body's equilibrium and homoeostasis. In considering these peptide amphiphiles, we can supplement our examination of bone regeneration with a presentation of the use of these materials in generating blood vessels, or angiogenesis. This is another important potential application of regenerative medicine, and we will return to this field at the end of this section, as it now constitutes a significant area of medical research in its own right.

One form that can be generated using peptide amphiphiles is that of a tube with inner hydrophobicity and an external functional moiety that can induce the binding and differentiation of the mesenchymal stem cells (MSCs) in the osteoblasts that synthesize bone matrix proteins and hydroxyapatite (the calcium and phosphate crystalline mineral base of bone). Using a sequence of peptides that attracts the Bone Morphogenic Protein (BMP-2)*, this amphiphilic ensemble is able to convert MSCs into the fully active osteoblasts that are responsible for the growth of the replacement bone. In order to exploit this property, researchers have designed submicrometric nanofibers that have been shown to induce effective bone regeneration by attracting BMP-2. Indeed, in an *in vivo* model, this kind of approach has shown the capacity to entirely fill the gap between two vertebrae.

Peptide amphiphiles can, like the graphene sheets we discussed earlier, be cross-linked and modified using the arginylglycylaspartic acid (RGD)* peptide—a well-established ligand that encourages cell adhesion—and phosphorylated serine to nucleate the mineralization of hydroxyapatite, thereby generating a structure similar to the three-dimensional structure of the original bone. These kinds of scaffolds have been seen to favor the process of mineralization, which can be prompted to begin in less than an hour. These synthesized peptide amphiphiles have even been mixed with implant precursors made using titanium, aluminum and vanadium, promoting mineralization (osteoconductivity) and osteoblast adhesion (biocompatibility) in experimental models, suggesting that these kinds of peptide amphiphiles are compatible with newly manufactured prostheses, opening up complementary avenues for their use. Thus, by harnessing the attachment of new bone, an appropriate peptide amphiphile could greatly improve the integration of the amalgam used in a prosthetic hip or knee with the patient's own tissue.

8.2.5 Hydrogels

There is also a lot of interest in hydrogels as a suitable scaffold material that can contribute to the construction of the extracellular matrix in bone restitution, combining a good biocompatibility profile with the possibility of degrading at a similar rate to the new bone formation (Yue et al., 2020). Hydrogels are networks of polymers that while being insoluble in water are capable of absorbing it, often in considerable quantities, to assume a large volume. In contact with water, they take it up to form semi-solid structures. The key property of these hydrogels in the context of bone regeneration is the stability of the three-dimensional polymers that make them up, as well as their biodegradability combined with a low risk of toxicity. The fact that these hydrogel structures also provide a favorable medium for delivering growth factors, and are hospitable for transplanted cells, makes it easier to regrow bone in this context (Bai et al., 2018).

The innovation at the nanotechnological level does not stop with these materials currently being studied in animal models. Complementary techniques add to their suitability, stimulating the replacement of missing tissue.

8.2.6 Magnetic Scaffolds

Techniques such as magnetic biohybrid porous scaffolds (crosslinking for collagen) employ magnetism to promote the regenerative processes. Here, introducing magnetic nanoparticles into a candidate combination of nanomaterials (such as a mix of apatite and collagen close to the composition of the natural bone matrix) has allowed researchers to use external magnetic fields to stimulate the regenerative capacity of these structures. In a form of the joule effect, a changing magnetic field applied from the outside affects the behavior of the stem cells, enhancing their regenerative potential. Similarly, magnetic nanoparticles have been shown to display regenerative properties in bone defects resulting from trauma or osteolysis. Polymeric scaffold using functionalized magnetic nanoparticles can be structured into three-dimensional networks allowing osteoblasts to induce magnetic-driven mineralization. Thus, the application of a weak magnetic field can induce osteoblast proliferation in a scaffold composed either of collagen or polylactide polymers, as well as hydroxyapatite combined with magnetic nanoparticles. While these magnetic scaffolds have been shown to be effective in the context of both in vitro and in vivo models, they still remain to be applied in human pathology (Tampieri et al., 2011).

But what of the driving force behind this regenerative medicine, the stem cells themselves? Unlike humans, lizards can grow back parts of their body, notably their tails, that have been lost in an accident, even though researchers have shown that the new tail is not quite the same as the one that was lost. Mastering the genetic triggers that allow this kind of regrowth will, it is hoped, allow humans to benefit from similar biological repair mechanisms. The discovery and use of stem cells in humans have opened the door, in principle at least, to re-growing or growing from scratch, any new organ or tissue from the patient herself or himself. Indeed, stem cells lie at the heart of regenerative medicine, so we need to take time to consider this area of research in order to better understand the functioning of these nanomaterials as hospitable material for the bone scaffold.

8.3 MESENCHYMAL STEM CELLS—THE KEY TO REGENERATIVE MEDICINE

MSCs are pluripotent hypoimmunogenic stromal cells isolated from different adult tissues including bone marrow, placenta or adipose tissue. One estimate is that less than one out of 10,000 cells is a MSC, but there are now well-established procedures for isolating them. Once placed in suitable conditions, these cells reproduce very quickly, allowing a good yield in only a few weeks of culture. Due to their ability to differentiate not only into bone marrow, adipocytes, chondrocytes, myocytes or osteocytes, but also into epithelial tissue, including neurons, hepatocytes or pancreatic cells, they were, as of early 2021, being tested in more than 1,200 clinical trials, although only six of these are associated with the use of nanoparticles.[2] Roughly half of these trials involve neurological, cardiovascular and bone and cartilage disease.

Due to their tropism for inflamed areas, one of the important properties of these MSCs is so-called homing, and they have the capacity to escape detection by the natural immune systems like leucocytes, and migrate to any area of inflammation. It is not surprising then that they are used in experimental models for transporting genes or nanoparticles into cancerous tumors. They are considered to be safer than standard untargeted nanoparticles as they minimize delivery of their therapeutic arsenal to non-cancerous cells. For the time being, MSC* transplantation is most frequently used to intervene in cardiac, neurological or cancerous pathologies. It is noteworthy that, according to the US Clinical Trial database, only about 30 studies have been completed as of early 2021 and more than 1,200 are still ongoing, with 47 for intestinal disease, 53 for cancers, 71 for skin disease, 76 for cardiac problems and 200 for diseases of the nervous system.[3]

The preponderance of trials in diseases of the central nervous system is due to another interesting property of these MSCs, as they have been found experimentally to have a tropism for injured nerve tissue. For this, they need to be loaded with magnetic nanoparticles, which induce plasma proteins that in turn help these stem cells to anchor onto the endothelial cells and to migrate across the blood-brain barrier in the same way that leucocytes do. Under certain conditions, these stem cells can attain areas of inflammation following traumatic brain injury and secrete anti-inflammatory factors (like TGF-beta) and neurotrophic factors promoting neurite outgrowth (the development of new neurons) and ultimately contributing to the

[2] https://clinicaltrials.gov, search using nanoparticle and mesenchymal as key words.

[3] As retrieved in January 2021 from US Clinical Trials database (https://clinicaltrials.gov/).

regeneration of nerves. These properties, which have already been observed using *in vitro* experiments, are now being further explored in the context of degenerative diseases of the central nervous system, such as lateral amyotrophic sclerosis or multiple sclerosis, where these MSCs have been shown to induce myelin production. In the case of amyotrophic lateral sclerosis (ALS or Charcot Disease), where MSC therapy has been shown to be effective in a rodent model, such autologous transplantation of MSC from bone marrow has been tested in human therapy and was shown to be effective in certain cases, as these transplanted MSC displayed similar immunomodulatory effects as seen in rodents and where they are associated with the clinical stabilization of ALS. While inconclusive, recent clinical trials hold out hope for the efficacy of MSC-based treatment of ALS (Cudkowicz et al., 2022). Similar therapy has been tested in spinal cord injury, where patients recovered their neurological function faster following the injection of several million MSCs into the area of the spinal lesion. In what follows, we briefly present some of the domains of medicine where stem cell therapies are currently being developed.

8.3.1 Diseases of the Bones and Joints

Osteoarthritis combines synovial inflammation, cartilage degeneration and subchondral osteofibrosis. The use of stem cells in a clinical trial of patients unresponsive to classic treatments of the condition has given positive results. The injection of MSCs into joints was followed by the alleviation of pain, increased joint mobility and the enhancement of the quality of the cartilage (Sharan et al., 2022). These experimental treatments need to be improved and stabilized perhaps with the use of targeted nanoparticles as vectors. To date (in 2021) only a handful have progressed to phase 3 studies, demonstrating that no specific treatment has yet emerged using these MSCs. But the experimental trials continue.

8.3.2 Heart Disease

A large number of preclinical and clinical phase 2 trials have shown some effectiveness of MSC transplantation in cases of heart disease and other injury, and recovery of cardiac function has been regularly noted following the intramyocardiac injection of these cells. The mechanism of action of these stem cells is not yet well enough understood, but probably involves the paracrine signaling system* between cells, with the secretion of growth factors that may activate or at least stimulate the resident cardiac stem cells of cardiomyocytes and the cells of the blood vessels.

However, looking overall at MSC therapy, we can note that the main clinical studies are still in the phase 1/2 stage and that while certain results are promising, real progress can only be expected following clarification of the mechanism of action of these stem cells. Whatever the disease and the area of treatment might be, researchers are still a long way from translating this approach from the experimental realm of the laboratory bench to its useful clinical application at the bedside. One area where applications seem to be ahead of the others is in the domain of angiogenesis.

8.4 ANGIOGENESIS

Angiogenesis is the formation of blood vessels in normal, cancerous or otherwise damaged tissue, an area that has also attracted considerable interest, particularly in the context of cancer research. This process is driven by growth factors like vascular endothelial growth factor (VEGF) and more generally those of the fibroblast growth factor (FGF) series: it enables tissue growth by supplying targeted sites with nutrients and oxygen. On the one hand, this can be largely beneficial in case of a regeneration project like the cases we were looking at above, and on the other hand, control over these processes can be used to impede the development of vascularization in case of abnormal or pathological tissue growth. Indeed, one of the reasons that there has been so much research on the process of angiogenesis is the hope of developing treatments against cancers that specifically target these mechanisms, cutting tumors off from the blood supply necessary for their growth. Of course, angiogenesis also holds out a certain hope for the treatment of heart disease, as many heart attacks result from problems with vascularization, although for the moment exploratory studies have instead pointed to the increased risk of vessels rupturing following attempts to stimulate vascularization.

To return to our examples of nanomaterials in this context, which ties us back into nanomedicine, we can note that synthetic peptide amphiphiles are able to induce angiogenesis either in their regular form or loaded with heparin fibroblast growth factor (FGF-2) or the vascular endothelial growth factor (VEGF). This induces the beta-sheet* conformation of complexes forming micrometric length structures of 7 nm in diameter. These nanofibers exhibit a high affinity for vascular growth factors, even leading to the formation of new vascularization in some animal models. The effective sequence of VEGF is a 15 amino acid length oligopeptide, and it displays the same neovascularization capability when grafted onto peptide amphiphiles, augmenting, for example, the proliferation of endothelial cells from the human umbilical vein (HUVEC).

Even though such applications are very interesting, applicable in animal models and open up new possibilities for both regenerative medicine and cancer therapy, there were no ongoing trials using these methods registered in 2021, at least none reported in the US Clinical Trials database.

8.5 THE POTENTIAL OF STEM CELLS

We have just seen how regenerative medicine embodies an interesting biomimetic approach to bone repair, a principle that can be expanded to any other organ in the body. This approach encourages renewed growth based on the elements that are already found in the host's body. Thus, benefitting from the multipotent cells already present, generally combined with growth-stimulating factors, the body's own repair mechanism can be harnessed and stimulated to make the bone regrow around the scaffold provided. But the better management of stem cells, and the discovery of new methods for stimulating even ordinary cells to recover their stem cell potential should open up new horizons, notably for autologous grafts. Here, we are talking not just about repair but also about replacement of tissue, and with the possibility of replacement comes the potential for enhancement.

8.6 FROM REPLACEMENT TO ENHANCEMENT

While for the moment regenerative medicine is principally oriented toward cure or compensation for lost tissue (and is, as we explained earlier, largely confined to preliminary laboratory studies), there is a thin line between curative replacement and enhancement. As in cosmetic surgery, the surgeon or the patient might want to profit from an intervention undertaken with the initial goal of restoring function to an organ in order to improve it, and I think most of us would find this quite legitimate, within reason. But who is to judge what is reasonable or justifiable in such an intervention? This issue of enhancement brings us onto the terrain of transhumanist philosophy.

Although its philosophical roots can be traced back as far as one would like, the transhumanist movement in its modern form dates from the closing decades of the twentieth century. The central tenet of the modern transhumanist movement is that we can use technology to transcend the human condition. And regenerative medicine looks like just the right kind of technology for doing this. When pre-emptively avoiding or treating disease, we are fully in the sphere of medicine, but the wish to improve the human being through the use of prostheses already promises to take us out of the bounds of consensual medical practice, at least in its classic conception as the art of curing infirmity and disease. The use of technology to replace defective or missing joints, limbs, organs or organelles opens two doors that interest the transhumanists. First, rather than just replacing failing or failed organs, it might be possible to replace these missing or defective organs with improved versions, and who would be opposed to this in principle? If you have the choice between a stronger or weaker upper arm, for example, who would choose the weaker one simply on the basis of not wanting to go beyond the natural capacities of a human limb (or even not wanting to go beyond the physical capacity of the patient before an automobile accident, for example)? The next step, then—and this is clearly outside the scope of normal medical practice—is to consider either supplementing or even replacing ordinary functioning organs with improved versions. Could certain humans be physically transcended by improved versions of themselves? Here, we cross into a world parallel to that of exoskeletons and other prosthetic mechanical enhancements of the human body that have attracted the interest of industrialists and the army alike. The ultimate possibility that goes beyond even enhancement is to replace used parts of the body in order to extend the patient's life, and once this process has been started, why not continue it indefinitely? In the spirit of the philosophical paradox of Theseus' ship*, there is potentially no limit to the renewal of a given individual. An unlimited capacity to replace tissues and organs, particularly if the 'spare parts' used for the replacement are generated from the patient's own body, opens the door to immortality, the grail of the transhumanists.

8.7 AUTOLOGOUS ORGAN TRANSPLANTATION

To conclude our consideration of regenerative medicine, we would like to back away from these heady speculations around immortality that are at the heart of transhumanist thought in order to underline one of the evident advantages of this approach if it can be successfully extended to organ transplantation. While the current methods are only capable of producing simple body tissues such as skin or bone, increasing mastery over the mechanisms of stem cell stimulation, growth and differentiation will enable researchers to produce more complex combinations of tissue, leading to the production of organs that can be used for transplantation. One of the biggest problems with organ transplantation in general is the compatibility between the donor and the recipient or the host. The matching of donors to recipients is a complex affair that occupies donor banks around the world. The artificial suppression of the host's immune system is an undesirable but necessary step for one of these transplants to work, and that, even after the painstaking process of matching donor to recipient as closely as possible. The patient, who is coming out of an already disruptive surgical operation, is left with substantially reduced resistance to infectious disease. Using the patient's own stem cells to grow a new organ around a nanomaterial scaffold would supply a perfect genetic match, removing the need for suppressing the immune response mechanism, and thereby increasing the chances of the long-term success of the procedure. A version of this approach is already used for bone marrow, with stem cells harvested from a patient prior to damaging chemotherapy before being reinjected post-chemotherapy to boost the production of blood cells after treatment. But more precise control over stem cells and the orientation of their growth could allow researchers to envisage using similar techniques (of harvesting and redeploying stem cells) in order to produce other organs for transplantation. On the model of the example of bone replacement elaborated earlier, similar contributions of nanomaterials could be made to structure the growth of such autologous organs and supply the conditions necessary for stimulating tissue growth. While the prospects of such autologous organ transplants still lie in the future, it is a field in which nanomaterials will be able to make a significant contribution to clinical outcomes.

8.8 TO GO FURTHER

Businaro, R., Corsi, M., Di Raimo, T., Marasco, S., Laskin, D. L., Salvati, B., Capoano, R., Ricci, S., Siciliano, C., Frati, G., & De Falco, E. (2016). Multidisciplinary approaches to stimulate wound healing. *Annals of the New York Academy of Sciences, 1378*(1), 137–142. https://doi.org/10.1111/nyas.13158

Kwon, S., Yoo, K. H., Sym, S. J., & Khang, D. (2019). Mesenchymal stem cell therapy assisted by nanotechnology: A possible combinational treatment for brain tumor and central nerve regeneration. *International Journal of Nanomedicine, 14*, 5925–5942. https://doi.org/10.2147/IJN.S217923

Patel, S., & Lee, K.-B. (2015). Probing stem cell behavior using nanoparticle-based approaches. *Wiley Interdisciplinary Reviews. Nanomedicine and Nanobiotechnology, 7*(6), 759–778. https://doi.org/10.1002/wnan.1346

Silva, L. H. A., Cruz, F. F., Morales, M. M., Weiss, D. J., & Rocco, P. R. M. (2017). Magnetic targeting as a strategy to enhance therapeutic effects of mesenchymal stromal cells. *Stem Cell Research & Therapy, 8*(1), 58. https://doi.org/10.1186/s13287-017-0523-4

Suh, H. (2000). Tissue restoration, tissue engineering and regenerative medicine. *Yonsei Medical Journal, 41*(6), 681–684. https://doi.org/10.3349/ymj.2000.41.6.681

Thomas, J., Thomas, S., Jiya, J., & Kalarikkal, N. (Eds.). (2017). *Recent trends in nanomedicine and tissue engineering*. River Publishers.

8.9 REFERENCES

Bai, X., Gao, M., Syed, S., Zhuang, J., Xu, X., & Zhang, X.-Q. (2018). Bioactive hydrogels for bone regeneration. *Bioactive Materials, 3*(4), 401–417. https://doi.org/10.1016/j.bioactmat.2018.05.006

Cudkowicz, M. E., Lindborg, S. R., Goyal, N. A., Miller, R. G., Burford, M. J., Berry, J. D., Nicholson, K. A., Mozaffar, T., Katz, J. S., Jenkins, L. J., Baloh, R. H., Lewis, R. A., Staff, N. P., Owegi, M. A., Berry, D. A., Gothelf, Y., Levy, Y. S., Aricha, R., Kern, R. Z., . . . Brown, R. H. (2022). A randomized placebo-controlled phase 3 study of mesenchymal stem cells induced to secrete high levels of neurotrophic factors in amyotrophic lateral sclerosis. *Muscle & Nerve, 65*(3), 291–302. https://doi.org/10.1002/mus.27472

Sharan, J., Barmada, A., Prodromos, C., & Candido, K. (2022). First human report of relief of lumbar and cervical discogenic and arthritic back pain after epidural and facet joint mesenchymal stem cell injection. *Current Stem Cell Research & Therapy*. https://doi.org/10.2174/1574888X17666220628123115

Tampieri, A., Landi, E., Valentini, F., Sandri, M., D'Alessandro, T., Dediu, V., & Marcacci, M. (2011). A conceptually new type of bio-hybrid scaffold for bone regeneration. *Nanotechnology, 22*(1), 015104. https://doi.org/10.1088/0957-4484/22/1/015104

Yue, S., He, H., Li, B., & Hou, T. (2020). Hydrogel as a biomaterial for bone tissue engineering: A review. *Nanomaterials, 10*(8), 1511. https://doi.org/10.3390/nano10081511

9 Conclusion

In this book, we have tried to give the reader a good idea of the current state of nanomedicine, exploring important areas of the field by means of illustrative examples. Given the growth of the domain since its inception at the end of the twentieth century, it would have been impossible to be exhaustive in our treatment, and so we chose sub-fields and examples that seemed particularly pertinent for limning this rapidly evolving domain. Using these examples, we have posed certain questions around the present state and the future of the burgeoning field of nanomedicine, tying it in to other movements in contemporary medical science, such as stem cell research and precision medicine. As we explained at the beginning, nanomedicine is, in principle, very wide-ranging, and the topics that we have covered in this book—nanopharmacy, theranostics, arrays and probes, genetic nanomedicine, organs on chips and regenerative medicine—although themselves quite different from each other, are far from exhausting what can be considered to fall under this heading. Nevertheless, the themes and the collection of examples used to illustrate the different areas provide a useful introduction to the field.

Apart from these chapters that are about the medical application of the nanosciences and nanotechnology proper, we have included a chapter on nanotoxicology. Although there have not been any proven cases of toxicity in humans associated with new nanoparticles issuing from research in the nano, we took the opportunity to present the reasons for thinking that certain nanomaterials or particles may well prove to be a threat to human health (Egbuna et al., 2021). Our aim here has not been to alarm the reader, but rather to explain why nanotoxicology deserves special attention and more significant funding than it currently receives. Already, the nanoparticles produced by diesel engines are suspected of having negative effects on human health, but, even here, turning statistical arguments around mortality in polluted environments into solid proof of a causal relationship between specific particles and human health problems is not an easy thing to do (Calderón-Garcidueñas et al., 2021).

The collection of examples of techniques that are being developed or deployed combined with some speculative thoughts about future uses of nanotechnology have allowed us to engage in wider discussions around nanomedicine, starting with its definition. A prominent philosophical problem with the nano, as with health, disease and medicine itself, is the difficulty in finding a suitable definition, one that is at the same time pertinent and practical to use, and one which allows us to envisage or conceptualize the field as a whole, even as it is developing. The choice to include all substances and objects, be they nanomaterials or other devices, that are less than 100 nm as the definition of the nano is practical, because it is precise and readily determined, but, as we have argued, it fails to reflect the breadth of the nano, or its core sense. In turn, while placing the nano in relation to a much wider project of miniaturization might make sense, this is not a good enough reason to propose a definition of the nano in terms of attempts at miniaturization, as this would provide a definition that

DOI: 10.1201/9781003367833-9

is much too broad, thereby losing touch with what is particularly significant about working at the level of the nanometer. So, while we have argued that there is a strong relationship between miniaturization projects and the nano, this does not mean that one should be equated with the other.

More telling in this debate over the definition of the nano is the case of objects that slip in and out of the category depending on their context, as we saw with the hepatitis C virus (HCV). While the evolution of the size and form of HCV as a function of its environment might help us better understand the shifting nature of nanoparticles in organic media, it once again shows up the artificiality of a 100 nm frontier between the nano and the non-nano. There may, however, be a possibility of finding a definition that takes this fluctuation in size into account. While working on definitions is one of the areas often invested by philosophy, we need to bear in mind that there are other domains for which a precise and minimally consensual definition of nanoparticles is important. Legislators, lawyers and financiers, for example, need a practical working definition of the nano that can answer the question: does this product or structure contain nanoparticles or not? In this context, the rule of 100 nm remains reasonably consensual and eminently practical. So, while it is part of the philosopher's job to point out any problems with, or more correctly the inconsistency of any definition, we are not foolhardy enough to propose an alternative definition of the nano or of nanomedicine.

Another philosophical, or more properly ethical, issue that we raised in the context of several different examples is that of the confidentiality of people's health information, particularly the digital health data generated by the nano within a patient. The general problem of the confidentiality of personal data is, of course, far from being limited to nanomedicine; the question of how to protect and share patients' digital data appropriately is one that applies to a whole range of digital information, from people's position on the planet (the GPS in your telephone) to their taste in music (on platforms like Spotify or Deezer). Some, although not all, patient data is particularly sensitive; information about the presence of a pre-cancerous condition, for example, could serve as a justification for increasing insurance premiums, if not refusing to provide health insurance altogether. There are, of course, laws, such as the General Data Protection Regulation (GDPR) in Europe, that protect personal data, but, while the protection of data that is stocked on a particular server is one thing, the protection of data that is available to anyone with the appropriate application on their mobile phone to detect it is quite another. Thus, a cause for particular concern with nanomedicine is that, as with the Internet of things, the digital data will no longer be concentrated in one place where it can be effectively regulated by the owner of the infrastructure under the supervision of public authorities. Instead, information supplied by nanotechnology will be disseminated throughout the patient population, depending for its interpretation on the techniques for 'reading' the data that risk being beyond the direct reach of legislative authorities, making it much more difficult to oversee (Bochud et al., 2022). Thinking about the ethical use of patient's health data inevitably takes us closer to the concerns of precision medicine, which is entirely based on the analysis of a patient's personal information (almost all in digital form) in order to determine the best available treatment for that person. Once again, while these techniques are confined to specialized hospital units, the institution can offer safeguards that protect the patient's personal health data. When the precision medicine approach has been disseminated into general practice, however, with the necessary tests undertaken in a substantial number of independent medical laboratories, the data is going to be harder to contain and control, perhaps even more so when the data is being managed by a private, for-profit company (Mittelstadt & Floridi, 2016).

Toward the end of the book, we have also questioned how nanomedicine will develop through the first half of the twenty-first century. It might seem paradoxical to conclude that while the nanomedicine discussed in this book has a bright future ahead of it, it is probably not, as we have suggested, in the form of a unified medical domain of nanomedicine. Instead, we have argued for the possibility that nanomedicine will dissolve or rather be disseminated into a set of sub-specialties within more traditional domains of medical specialization: nano-nephrology, nano-oncology, reproductive nano-medicine, etc. While this might trigger the decline of the use of nanomedicine as a keyword or category for classification, it would not undo the contributions that have been made under this head, which will continue to develop in other disciplinary niches.

The ideas and associated research that are driving nanomedicine are impressive, in terms of both their ingenuity and their number. New publications appear all the time, generally touting targeted, highly specific treatments, potentially linked to diagnostic tools that involve deploying the same or similar nanoparticles. Nanoscience and nanotechnology have the potential to transform medicine, just as they have the potential to change many other areas of life.

This being said, it has now been over 20 years since the first nano-based treatments were approved for use, and we cannot say that the nano have as yet transformed medicine, not even in the more intensively explored area of oncology. More and more vectorized products are available, for example, but they have not yet superseded older chemotherapy. While we have mentioned the contribution of nanoparticles such as silver and gold to disinfection and imaging, we have maybe underestimated the importance of these applications in the practical introduction of nanoparticles into medical treatment (Deshmukh et al., 2019). The temptation is always to look to more sophisticated and innovative uses of the nano, such as in regenerative medicine and theranostics, but these metal nanoparticles may well prove an invaluable new tool for combatting drug-resistant infectious disease, a very real problem in contemporary medicine (Huang et al., 2022).

We can close, then, by returning to the original question in the title of the book: should nanomedicine be viewed as a potential panacea, or as Pandora's box? Which describes it better? The universal cure or an unnecessary risk? While both seem excessive, with time we will better be able to place the cursor between them. On the positive side, we have seen a range of applications that are still in development but promise to improve the handling of a number of diseases, notably cancers, but also the chronic and neurological diseases that affect our ageing populations. But what about the negative side? For nanomedicine in particular, the threat of abusive or even illegal enhancement still seems a long way off (even if some transhumanists see it already on the horizon); on the other hand, the clinical applications of nanomedicine are yet to transform medical practice. But even the failure of all the examples we have presented in this book to provide innovations that would help in the identification and treatment of disease—an outcome that seems very unlikely—would not turn nanomedicine into the scourge of human health; it would just be disappointing and involve a considerable waste of resources. Thus, in terms of nano-based drugs, there are enough precautions in place, particularly on the side of clinical trials that we are unlikely to see a highly toxic product put into clinical use, let alone disseminated around the planet.

The risk that would merit the qualification of Pandora's box would be the introduction of harmful man-made nanoparticles or nanomaterials into the environment. By describing the particularities of the action of nanoparticles in the body, the chapter on nanotoxicity was meant to show how and why this could conceivably come about, and outline what should be done to avoid it. Here, the answer is to increase investment in nanotoxicological studies covering not only new nanomaterials that will be put on the market but also the increasing number of combinations of nanoparticles that will be found in our environment.

Nanomedicine, whatever its future form, will finish by integrating our lives. A number of the projects that we have presented here will be part of it, along with a larger number of initiatives that we have not discussed, due to lack of space, or simply because they have yet to be launched. Paradoxically, we see nanomedicine as still to a large degree lying in the future, and yet, as we have argued, it is already dissolving into branches that correspond to the application of nano-techniques within better established medical specialties. We feel confident that nanomedicine has a bright future ahead of it, and will, in the end, transform medical practice, maybe not immediately, maybe not radically, but profoundly and permanently.

9.1 REFERENCES

Bochud, M., Le Pogam, M.-A., Thabard, J., & Monod, S. (2022). Données de santé: Le nouvel or numérique, mais pour qui? [Health data: The new digital gold, but for whom?]. *Revue Médicale Suisse, 18*(790), 1406–1411. https://doi.org/10.53738/REVMED.2022.18.790.1406

Calderón-Garcidueñas, L., Stommel, E. W., Rajkumar, R. P., Mukherjee, P. S., & Ayala, A. (2021). Particulate air pollution and risk of neuropsychiatric outcomes. What we breathe, swallow, and put on our skin matters. *International Journal of Environmental Research and Public Health, 18*(21), 11568. https://doi.org/10.3390/ijerph182111568

Deshmukh, S. P., Patil, S. M., Mullani, S. B., & Delekar, S. D. (2019). Silver nanoparticles as an effective disinfectant: A review. *Materials Science and Engineering: C, 97*, 954–965. https://doi.org/10.1016/j.msec.2018.12.102

Egbuna, C., Parmar, V. K., Jeevanandam, J., Ezzat, S. M., Patrick-Iwuanyanwu, K. C., Adetunji, C. O., Khan, J., Onyeike, E. N., Uche, C. Z., Akram, M., Ibrahim, M. S., El Mahdy, N. M., Awuchi, C. G., Saravanan, K., Tijjani, H., Odoh, U. E., Messaoudi, M., Ifemeje, J. C., Olisah, M. C., . . . Ibeabuchi, C. G. (2021). Toxicity of nanoparticles in biomedical application: Nanotoxicology. *Journal of Toxicology, 2021*, 9954443. https://doi.org/10.1155/2021/9954443

Huang, Y.-S., Wang, J.-T., Tai, H.-M., Chang, P.-C., Huang, H.-C., & Yang, P.-C. (2022). Metal nanoparticles and nanoparticle composites are effective against *Haemophilus influenzae, Streptococcus pneumoniae*, and multidrug-resistant bacteria. *Journal of Microbiology, Immunology, and Infection = Wei Mian Yu Gan Ran Za Zhi*, S1684–1182(22)00071–8. https://doi.org/10.1016/j.jmii.2022.05.003

Mittelstadt, B. D., & Floridi, L. (2016). The ethics of big data: Current and foreseeable issues in biomedical contexts. *Science and Engineering Ethics, 22*(2), 303–341. https://doi.org/10.1007/s11948-015-9652-

Glossary

The definitions were retrieved from the Medical Subject Heading thesaurus (MeSH browser: https://meshb.nlm.nih.gov/search). The date between brackets was the entry date of the descriptor. When no date is mentioned, the definition was adapted from Wikipedia. Some definitions of descriptors have been adapted or truncated.

Abraxane®: An injectable formulation of albumin-bound paclitaxel nanoparticles (2016/01/01).

Adenosine triphosphate (ATP): An organic compound and hydrotrope that provides energy to drive many processes in living cells, such as muscle contraction, nerve impulse propagation, condensate dissolution and chemical synthesis. Found in all known forms of life, ATP is often referred to as the 'molecular unit of currency' of intracellular energy transfer.

Adenovirus: A family of non-enveloped viruses infecting mammals (Mastadenovirus) and birds (Aviadenovirus) or both (Atadenovirus). Infections may be asymptomatic or result in a variety of diseases (1981/01/01).

Adipocyte: Cell in the body that stores fats, usually in the form of triglycerides. White adipocytes are the predominant type and found mostly in the abdominal cavity and subcutaneous tissue. Brown adipocytes are thermogenic cells that can be found in newborns of some species and hibernating mammal (1994/01/01).

Aerodynamic diameter: The aerodynamic diameter of an irregular particle is defined as the diameter of the spherical particle with a density of 1000 kg/m^3 and the same settling velocity as the irregular particle.

Age-related macular degeneration (ARMD): Degenerative changes in the retina usually of older adults which result in a loss of vision in the center of the visual field (the macula lutea) because of damage to the retina. It occurs in dry and wet forms (1979/01/01).

AIDS (acquired immune deficiency syndrome): A spectrum of conditions caused by infection with the human immunodeficiency virus (HIV), a retrovirus. Following the initial infection an individual may not notice any symptoms, or may experience a brief period of influenza-like illness. Typically, this is followed by a prolonged incubation period with no symptoms. If the infection progresses, it interferes more with the immune system, increasing the risk of developing common infections such as tuberculosis, as well as other opportunistic infections, and tumors which are otherwise rare in people who have normal immune function. Since the 1990s, tritherapy (a combination of three drugs taken together) has been used to successfully control the HIV and so avoid AIDS, but there is, for the moment, no known cure for HIV.

Albumins: Water-soluble proteins found in egg whites, blood, lymph, and other tissues and fluids. They coagulate upon heating (1966/01/01).

Alginate: Salts and esters of alginic acid that are used as hydrogels, dental impression materials and as absorbent materials for surgical dressings (bandages, hydrocolloid). They are also used to manufacture microspheres and nanoparticles for diagnostic reagent kits and drug delivery systems (1966/01/01).

Alveoli-capillary barrier or blood-Air barrier: The barrier between capillary blood and alveolar air comprising the alveolar epithelium and capillary endothelium with their adherent basement membrane and epithelial cell cytoplasm. Pulmonary gas exchange occurs across this membrane (1990/01/01).

Alzheimer disease: A degenerative disease of the brain characterized by the insidious onset of dementia. Impairment of memory, judgment, attention span and problem-solving skills are followed by severe apraxias and a global loss of cognitive abilities. The condition primarily occurs after age 60, and is marked pathologically by severe cortical atrophy and the triad of senile plaques; neurofibrillary tangles and neuropil threads (1984/01/01). (From Adams et al., *Principles of Neurology, 6th ed.*, pp. 1049–57.)

Amyloidosis: A group of sporadic, familial and/or inherited, degenerative and infectious disease processes, linked by the common theme of abnormal protein folding and deposition of amyloid. As the amyloid deposits enlarge, they displace normal tissue structures, causing disruption of function. Various signs and symptoms depend on the location and size of the deposits (1966/01/01).

Amyotrophic lateral sclerosis: A degenerative disorder affecting upper motor neurons in the brain and lower motor neurons in the brain stem and spinal cord. Disease onset is usually after the age of 50 and the process is usually fatal within three to six years. Clinical manifestations include progressive weakness, atrophy, fasciculation, hyperreflexia, dysarthria, dysphagia and eventual paralysis of respiratory function. Pathologic features include the replacement of motor neurons with fibrous astrocytes and atrophy of anterior spinal nerve roots and corticospinal tracts (1966/01/01). (From Adams et al., *Principles of Neurology, 6th ed.*, pp. 1089–94.)

Anamnesis: 'The information gained by a physician about a patient medical story (from Greek: ἀνά, aná, "entire", and μνῆσις, mnesis, "memory")'

Anaphylactic shock: An acute hypersensitivity reaction due to exposure to a previously encountered antigen. The reaction may include rapidly progressing urticaria, respiratory distress, vascular collapse, systemic shock and death (1966/01/01).

Anatoxins or toxoids: Preparations of pathogenic organisms or their derivatives made nontoxic and intended for active immunologic prophylaxis. They include deactivated toxins (1966/01/01).

Angiogenesis: Or neovascularization pathologic. A pathologic process consisting of the proliferation of blood vessels in abnormal tissues or in abnormal positions (1980/01/01).

Aniline: Organic compound consisting of a phenyl group attached to an amino group.

Anthracycline: Compound extracted from Streptomyces bacterium with four linked phenyl groups and a sugar moiety.

Antigen: Substances that are recognized by the immune system and induce an immune reaction (1966/01/01).

Apolipoprotein E: Protein components on the surface of lipoproteins. They form a layer surrounding the hydrophobic lipid core. There are several classes of apolipoproteins with each playing a different role in lipid transport and lipid metabolism. These proteins are synthesized mainly in the liver and the intestines (1977/01/01).

Apoptosis: A regulated cell death mechanism characterized by distinctive morphologic changes in the nucleus and cytoplasm, including the endonucleolytic cleavage of genomic DNA, at regularly spaced, internucleosomal sites, that is, DNA fragmentation. It is genetically programmed and serves as a balance to mitosis in regulating the size of animal tissues and in mediating pathological processes associated with tumor growth (1993/01/01).

Aptamer peptides: Peptide sequences, generated by iterative rounds of SELEX aptamer technique, that bind to a target molecule specifically and with high affinity (2006/01/01).

ARMD: Age-related macular degeneration.

aRNA: Or linear amplified RNA refers to an in vitro RNA transcription-mediated amplification according to LI, J., Eberwine, J. The successes and future prospects of the linear antisense RNA amplification methodology. Nat Protoc 13, 811–818 (2018).

Asbestos: Fibrous incombustible mineral composed of magnesium and calcium silicates with or without other elements. It is relatively inert chemically and used in thermal insulation and fireproofing. Inhalation of dust causes asbestosis and later lung and gastrointestinal neoplasms (1970/01/01).

Astrocyte: A class of large neuroglial (macroglial) cells in the central nervous system—the largest and most numerous neuroglial cells in the brain and spinal cord. Astrocytes (from 'star' cells) are irregularly shaped with many long processes, including those with 'end feet', which form the glial (limiting) membrane and directly and indirectly contribute to the blood-brain barrier. They regulate the extracellular ionic and chemical environment, and 'reactive astrocytes' (along with microglia) respond to injury (1979/01/01).

ATP: Adenosine triphosphate.

Autograft: Transplant comprising an individual's own tissue, transferred from one part of the body to another (2014/01/01).

Autophagy: The segregation and degradation of various cytoplasmic constituents via engulfment by multivesicular bodies, vacuoles, or autophagosomes and their digestion by lysosomes. It plays an important role in biological metamorphosis and in the removal of bone by osteoclasts. Defective autophagy is associated with various diseases, including neurodegenerative diseases and cancer (1991/01/01).

B-lymphocytes: Lymphoid cells concerned with humoral immunity. They are short-lived cells resembling bursa-derived lymphocytes of birds in their production of immunoglobulin upon appropriate stimulation (1973/01/01).

***Bacillus anthracis*:** A species of bacteria that causes anthrax in humans and animals (1966/01/01).

Beta sheet: A common motif of the regular protein secondary structure and folding.

Beta-adrenoreceptors: Also called Receptors, Adrenergic, beta. One of two major pharmacologically defined classes of adrenergic receptors. The beta adrenergic receptors play an important role in regulating cardiac muscle contraction, smooth muscle relaxation and glycogenolysis (1984/01/01).

Beta-blockers: See Propranolol for an example of beta-blocker.

Biocompatibility: See Biocompatible Materials.

Biocompatible materials: Synthetic or natural materials, other than drugs, that are used to replace or repair any body tissues or bodily function (1973/01/01).

Biofilms: Encrustations formed from microbes (bacteria, algae, fungi, plankton or protozoa) embedded in an extracellular polymeric substance matrix that is secreted by the microbes. They occur on body surfaces such as teeth (dental deposits), inanimate objects and bodies of water. Biofilms are prevented from forming by treating surfaces with dentifrices, disinfectants, anti-infective agents and anti-fouling agents (1995/01/01).

Biomarker: Measurable and quantifiable biological parameters (e.g., specific enzyme concentration, specific hormone concentration, specific gene phenotype distribution in a population, presence of biological substances) which serve as indices for health- and physiology-related assessments, such as disease risk, psychiatric disorders, environmental exposure and its effects, disease diagnosis; metabolic processes; substance abuse; pregnancy; cell line development; epidemiological studies; etc. (1989/01/01)

Biopersistency: Persistency of xenobiotics in an organism.

Biopsy: Removal and pathologic examination of specimens from the living body (1966/01/01).

BMP-2: Bone morphogenic protein type 2.

Bone marrow stromal cells: See Mesenchymal Stem Cells.

Bone morphogenetic proteins: Bone-growth regulatory factors that are members of the transforming growth factor-beta superfamily of proteins. They are synthesized as large precursor molecules, which are cleaved by proteolytic enzymes. The active form can consist of a dimer of two identical proteins or a heterodimer of two related bone morphogenetic proteins (1997).

Boyden chamber: Chambers isolated by filters are the appropriate tools for the accurate determination of cells' capacity for migration behavior.

Brownian diffusion/motion or pedesis: The random motion of particles suspended in a medium.

Buckminster fullerene: See Fullerene.

Carcinogenic potential: See Carcinogens.

Carcinogens: Substances that increase the risk of neoplasms in humans or animals. Both genotoxic chemicals, which affect DNA directly, and nongenotoxic chemicals, which induce neoplasms by other mechanism, are included (1966/01/01).

Cardiomyocyte: Striated muscle cells found in the heart. They are derived from cardiac myoblasts (2003/01/01).

Cardiotonic drugs: Agents that have a strengthening effect on the heart or that can increase cardiac output. They may be cardiac glycosides, sympathomimetics or other drugs. They are used after myocardial infarct, cardiac surgical procedures, in shock or in congestive heart failure (1980/01/01).

Cation or cationic ion: A positively charged atoms, radicals or groups of atoms, which travel to the cathode or negative pole during electrolysis (1978/01/01).

Caveolin: The main structural proteins of caveolae. Several distinct genes for caveolins have been identified (2001/01/01).

CD47 (antigen): A ubiquitously expressed membrane glycoprotein. It interacts with a variety of integrins and mediates responses to extracellular matrix proteins (2006/01/01).

cDNA: Single-stranded complementary DNA synthesized from an RNA template by the action of RNA-dependent DNA polymerase. cDNA (i.e., complementary DNA, not circular DNA, not c-DNA) is used in a variety of molecular cloning experiments as well as serves as a specific hybridization probe (1994/01/01).

Chemotherapy (drug therapy): The use of drugs to treat a disease or its symptoms. One example is the use of antineoplastic agents to treat cancer (1966/01/01).

Chloramine: Inorganic derivatives of ammonia by substitution of one or more hydrogen atoms with chlorine atoms or organic compounds with the general formulas R2nCl and RnCl2 (where R is an organic group) (1966/01/01).

Cholesterol: The principal sterol of all higher animals, distributed in body tissues, especially the brain and spinal cord, and in animal fats and oils (1960/01/01).

Chondrocyte: Polymorphic cells that form cartilage (1998/01/01).

Chromatography: Techniques used to separate mixtures of substances based on differences in the relative affinities of the substances for mobile and stationary phases. A mobile phase (fluid or gas) passes through a column containing a stationary phase of porous solid or liquid coated on a solid support. Usage is both analytical for small amounts and preparative for bulk amounts (1966/01/01).

Chronic obstructive pulmonary disease: A disease of chronic diffuse irreversible airflow obstruction. Subcategories of COPD include chronic bronchitis and pulmonary emphysema (2002/01/01).

Chylomicrons: A class of lipoproteins that carry dietary cholesterol and triglycerides from the small intestine to the tissues. Their density (0.93–1.006 g/ml) is the same as that of very-low-density lipoproteins (1966/01/01).

Clathrin: The main structural coat protein of coated vesicles which play a key role in the intracellular transport between membranous organelles. Each molecule of clathrin consists of three light chains (clathrin light chains) and three heavy chains (clathrin heavy chains) that form a structure called a triskelion. Clathrin also interacts with cytoskeletal proteins (1984/01/01).

CNT: Carbon nanotube.

Collagen: A polypeptide substance comprising about one-third of the total protein in mammalian organisms. It is the main constituent of skin, connective tissue, and the organic substance of bones and teeth (1966/01/01).

Confocal microscopy: A light microscopic technique in which only a small spot is illuminated and observed at a time. An image is constructed through point-by-point scanning of the field in this manner. Light sources may be conventional or laser, and fluorescence or transmitted observations are possible (1995/01/01).

COPD: Chronic obstructive pulmonary disease.

Coronome: A subset of serum proteins that are bound in the 'corona' of a nanoparticle that gives a specific signature of attached protein (Rihn BH and Joubert O, ACS Nano, 2015, 9 (6), pp. 5634–5635 DOI: 10.1021/acsnano.5b00459). It should not be confused with the corona form of the Coronavirus.

COVID-19: Coronavirus disease that appeared in 2019, leading to the pandemic of 2020–2021.

Creutzfeldt-jakob syndrome: A rare transmissible encephalopathy most prevalent between the ages of 50 and 70. Affected individuals may present with sleep disturbances, personality changes and ataxia; aphasia, visual loss, weakness, muscle atrophy, myoclonus, progressive dementia and death within one year of disease onset. A familial form exhibiting autosomal dominant inheritance and a new variant CJD (potentially associated with encephalopathy, bovine spongiform) has been described. Pathological features include prominent cerebellar and cerebral cortical spongiform degeneration and the presence of prions (1973/01/01). (From N Engl J Med, 1998 Dec 31; 339(27).)

CTC: Circulating Tumor Cells.

Cytostatic agents: Compounds that inhibit or prevent the proliferation of cells (2008/01/01).

Dendritic cell: Specialized cells of the hematopoietic system that have branch-like extensions. They are found throughout the lymphatic system, and in non-lymphoid tissues such as skin and the epithelia of the intestinal, respiratory and reproductive tracts. They trap and process antigens, and present them to T-cells, thereby stimulating cell-mediated immunity (1987/01/01).

Dermis: A layer of vascularized connective tissue underneath the epidermis. The surface of the dermis contains innervated papillae. Embedded in or beneath the dermis are sweat glands, hair follicles and sebaceous glands (1999/01/01).

Disease stratification: A disease mechanism-based patient stratification at the molecular level to address the needs for stratified or personalized therapeutic interventions (source Cordis, EU).

DNA damage: Injuries to DNA that introduce deviations from its normal, intact structure and which may, if left unrepaired, result in a mutation or a block of DNA replication. These deviations may be caused by physical or chemical agents and occur by natural or unnatural, introduced circumstances. They include the introduction of illegitimate bases during replication or by deamination or other modification of bases; the loss of a base from the DNA backbone leaving an abasic site; single-stranded breaks; double-stranded breaks; and intrastrand (pyrimidine dimers) or interstrand crosslinking. Damage can often be repaired (DNA repair). If the damage is extensive, it can induce apoptosis (1987/01/01).

DNA mismatch repair: A DNA repair pathway involved in correction of errors introduced during DNA replication when an incorrect base, which cannot form hydrogen bonds with the corresponding base in the parent strand, is incorporated into the daughter strand. Excinucleases recognize the base pair mismatch and cause a segment of polynucleotide chain to be excised from the daughter strand, thereby removing the mismatched base (from *Oxford Dictionary of Biochemistry and Molecular Biology*, 2001), (2007/01/01).

DNA polymerase: DNA-dependent DNA polymerases found in bacteria, animal and plant cells. During the replication process, these enzymes catalyze the addition of deoxyribonucleotide residues to the end of a DNA strand in the presence of DNA as template-primer (1977/01/01).

DNA replication: The process by which a DNA molecule is duplicated (1968/01/01).

DNA-directed RNA polymerases: Enzymes that catalyze DNA template-directed extension of the 3′-end of an RNA strand one nucleotide at a time. They can initiate a chain de novo. In eukaryotes, three forms of the enzyme have been distinguished on the basis of sensitivity to alpha-amanitin, and the type of RNA synthesized (From Enzyme Nomenclature, 1992) (1973/01/01).

Doxorubicin: Anthracycline derivative that is antineoplastic by intercalating between base pairs of DNA.

Ebola virus: A genus in the family filoviridae consisting of several distinct species of eBola virus, each containing separate strains. These viruses cause outbreaks of a contagious, hemorrhagic disease (hemorrhagic fever, Ebola) in humans, usually with high mortality (2002/01/01).

ECM: See Extracellular Matrix.

Elastomer or elastic polymer: A generic term for all substances having the properties of stretching under tension, high tensile strength, retracting rapidly and recovering their original dimensions fully. They are generally polymers (1999/11/03).

Electrophoresis: An electrochemical process in which macromolecules or colloidal particles with a net electric charge migrate in a solution under the influence of an electric current (1966/01/01).

EMA: European Medicines Agency.

Encoded protein: A protein coded by a gene via a transcription by an mRNA.

Endocytosis: Cellular uptake of extracellular materials within membrane-limited vacuoles or microvesicles. Endosomes play a central role in endocytosis (1976/01/01).

Endosteal niche: A part of the stem cell niche, that is a particular zone of tissue composed of a specialized microenvironment where stem cells are retained in an undifferentiated, self-renewable state (2009/01/01).

Endothelial cells: Highly specialized epithelial cells that line the heart; blood vessels and lymph vessels, forming the endothelium. They are polygonal in shape and joined together by tight junctions. The tight junctions allow for variable permeability to specific macromolecules that are transported across the endothelial layer (2004/01/01).

Enzyme: Biological molecules that possess catalytic activity. They may occur naturally or be synthetically created. Enzymes are usually proteins, however catalytic RNA and catalytic DNA molecules have also been identified (1966/01/01).

Epigenetics or epigenomics: The systematic study of the global gene expression changes due to epigenetic processes and not due to DNA base sequence changes (2011/01/01).

Epitope: Sites on an antigen that interact with specific antibodies (1972/01/01).

ERV: Endogenous retrovirus.

Estrogens: Compounds that interact with estrogen receptors in target tissues to bring about the effects similar to those of estradiol. Estrogens stimulate the female reproductive organs, and the development of secondary female sex characteristics. Estrogenic chemicals include natural, synthetic, steroidal or non-steroidal compounds (1966/01/01).

ETH: *Eidgenössische Technische Hochschule* in Zürich, Switzerland.

Ethmoid bone: A light and spongy (pneumatized) bone that lies between the orbital part of frontal bone and the anterior of sphenoid bone. Ethmoid bone separates the orbit from the ethmoid sinus. It consists of a horizontal plate, a perpendicular plate and two lateral labyrinths (1960/01/01).

Evidence-based medicine (EBM): An approach of practicing medicine with the goal to improve and evaluate patient care. It requires the judicious integration of best research evidence with the patient's values to make decisions about medical care. This method is to help physicians make proper diagnosis, devise best testing plan, choose best treatment and methods of disease prevention as well as develop guidelines for large groups of patients with the same disease (from *JAMA* 296/9, 2006) (1997/01/01).

Exposome: The measure of all the exposures of an individual from all sources, including environmental and occupational sources, in a lifetime and how those exposures relate to health (2020/01/01) (from www.cdc.gov/niosh/topics/exposome/on 06/06/2019).

Extracellular matrix: A meshwork-like substance found within the extracellular space and in association with the basement membrane of the cell surface. It promotes cellular proliferation and provides a supporting structure to which cells or cell lysates in culture dishes adhere (1984/01/01).

Extravasation: Extravasation is the leakage of a fluid out of its container into the surrounding area, especially blood or blood cells from vessels.

FDA: Food and Drug Administration (United States regulatory agency).

Ferromagnetism: Ferromagnetism is the basic mechanism by which certain materials (such as iron) form permanent magnets, or are attracted to magnets.

FGF: See Fibroblast Growth Factor.

Fibroblast: Connective tissue cell which secrete an extracellular matrix rich in collagen and other macromolecules (1965/01/01).

Fibroblast growth factor: A family of small polypeptide growth factors that share several common features including a strong affinity for heparin, and a central barrel-shaped core region of 140 amino acids that is highly homologous between family members. Although originally studied as proteins that stimulate the growth of fibroblasts, this distinction is no longer a requirement for membership in the fibroblast growth factor family (1984/01/01).

Fluorescent dye: Chemicals that emit light after excitation by light. The wavelength of the emitted light is usually longer than that of the incident light. Fluorochromes are substances that cause fluorescence in other substances, that is, dyes used to mark or label other compounds with fluorescent tags (1966/01/01).

Folate: The active part of Folic Acid.

Folic acid: A member of the vitamin B family that stimulates the hematopoietic system. It is present in the liver and kidney and is found in mushrooms, spinach, yeast, green leaves and grasses (*Poaceae*). Folic acid is used in the treatment and prevention of folate deficiencies and megaloblastic anemia (1960/01/01).

Fullerene: A polyhedral carbon structure composed of around 60–80 carbon atoms in pentagon and hexagon configuration. They are named after Buckminster Fuller because of structural resemblance to geodesic domes. Fullerenes can be made in high temperature such as arc discharge in an inert atmosphere (2003/01/01).

Gastrin: Member of a family of gastrointestinal peptide hormones that excite the secretion of gastric juice. They may also occur in the central nervous system where they are presumed to be neurotransmitters (1978/01/01).

Genetic engineering: Or genetic manipulation, directed modification of the gene complement of a living organism by such techniques as altering the DNA, substituting genetic material by means of a virus, transplanting whole nuclei, transplanting cell hybrids, etc. (1973/01/01)

Genomics: The systematic study of the complete DNA sequences (genome) of organisms. Included are construction of complete genetic, physical and transcript maps, and the analysis of this structural genomic information on a global scale such as in genome wide association studies (2001/01/01).

Genotoxicity: The study of the ability of chemical or physical agents to induce DNA damage. See DNA Damage.

Genotype: The genetic constitution of the individual, comprising the alleles present at each genetic locus (1968/01/01).

GEO: Gene Expression Omnibus (www.ncbi.nlm.nih.gov/geo/browse/).

Glioblastoma: A malignant form of astrocytoma histologically characterized by pleomorphism of cells, nuclear atypia, microhemorrhage and necrosis. They may arise in any region of the central nervous system, with a predilection for the cerebral hemispheres, basal ganglia and commissural pathways. Clinical presentation most frequently occurs in the fifth or sixth decade of life with focal neurologic signs or seizures (1994/01/01).

Glycerolipids: Glycerolipids are composed of mono-, di-, and tri-substituted glycerols, the best-known being the fatty acid triesters of glycerol, called triglycerides.

Graphene: An allotropic form of carbon that is used in pencils, as a lubricant, and in matches and explosives. It is obtained by mining and its dust can cause lung irritation (1991/01/01). Also widely used in nanomaterials.

Growth factor: Regulatory proteins and peptides that are signaling molecules involved in the process of paracrine communication. They are generally considered factors that are expressed by one cell and are responded to by receptors on another nearby cell. They are distinguished from hormones in that their actions are local rather than distal (2003/01/01).

Half-life: The time it takes for a substance (drug, radioactive nuclide or other) to lose half of its pharmacologic, physiologic or radiologic activity (1974/01/01).

HCG: See Human Chorionic Gonadotropin.

HCV: Hepatitis C virus.

Hemagglutinin: Agents that cause agglutination of red blood cells. They include antibodies, blood group antigens, lectins, autoimmune factors, bacterial, viral or parasitic blood agglutinins (1981/01/01).

Hematopoietic stem cells: Progenitor cells from which all blood cells derived. They are found primarily in the bone marrow and also in small numbers in the peripheral blood (1972/01/01).

Hepatic portal system: A system of vessels in which blood, after passing through one capillary bed, is conveyed through a second set of capillaries before it returns to the systemic circulation (1966/01/01). It allows the transport of venous blood from spleen, pancreas, gallbladder and the abdominal portion of the gastrointestinal tract into the liver.

Hepatic stellate cell: Perisinusoidal cells of the liver, located in the space of Disse between hepatocytes and sinusoidal endothelial cells (2009/01/01).

Hepatocyte: The main structural component of the liver. They are specialized epithelial cells that are organized into interconnected plates called lobules (2001/01/01).

HER 2: Human Epidermal Growth-Factor Receptor.

Herceptin®: See Trastuzumab.

High-throughput nucleotide sequencing: Techniques of nucleotide sequence analysis that increase the range, complexity, sensitivity and accuracy of results by greatly increasing the scale of operations and thus the number of nucleotides, and the number of copies of each nucleotide sequenced. The sequencing may be done by analysis of the synthesis or ligation products, hybridization to pre-existing sequences, etc. (2011/01/01)

Homeostasis: The processes whereby the internal environment of an organism tends to remain balanced and stable (1966/01/01).

Hormesis: Biphasic dose responses of cells or organisms (including microorganisms) to an exogenous or intrinsic factor, in which the factor induces stimulatory or beneficial effects at low doses and inhibitory or adverse effects at high doses (2012/01/01).

Hormones: Chemical substances having a specific regulatory effect on the activity of a certain organ or organs. The term was originally applied to substances secreted by various endocrine glands and transported in the bloodstream to the target organs. It is sometimes extended to include those substances that are not produced by the endocrine glands but that have similar effects (1966/01/01).

HT(N)S: High-Throughput (Nucleotide) Sequencing.

HUGO: Human Genome Project.

Human chorionic gonadotropin: A gonadotropic glycoprotein hormone produced primarily by the placenta. Similar to the pituitary luteinizing hormone in structure and function, chorionic gonadotropin is involved in maintaining the corpus luteum during pregnancy. hCG consists of two noncovalently linked subunits, alpha and beta. Within a species, the alpha subunit is virtually identical to the alpha subunits of the three pituitary glycoprotein hormones (TSH, LH, and FSH), but the beta subunit is unique and confers its biological specificity (2003/01/01). This hormone is dosed to confirm pregnancy.

Hyaluronic acid or hyaluronate: A natural high-viscosity mucopolysaccharide with alternating beta (1–3) glucuronide and beta (1–4) glucosaminidic bonds. It is found in the umbilical cord, in vitreous body and in synovial fluid. A high urinary level is found in Progeria (2002/01/01).

Hybridization: Widely used technique which exploits the ability of complementary sequences in single-stranded DNA or RNA to pair with each other to form a double helix. Hybridization can take place between two complementary DNA sequences, between a single-stranded DNA and a complementary RNA, or between two RNA sequences. The technique is used to detect and isolate specific sequences, measure homology or define other characteristics of one or both strands (Kendrew, *Encyclopedia of Molecular Biology*, 1994, p503) (1972/01/01).

Hydrogel: Water swollen, rigid, three-dimensional network of cross-linked, hydrophilic macromolecules, 20%–95% water. They are used in paints, printing inks, foodstuffs, pharmaceuticals and cosmetics (1999/01/01).

Hydrogen peroxide: A strong oxidizing agent used in aqueous solution as a ripening agent, bleach and topical anti-infective. It is relatively unstable and solutions deteriorate over time unless stabilized by the addition of acetanilide or similar organic materials (1966/01/01).

Hydrolysis: The process of cleaving a chemical compound by the addition of a molecule of water (1991/01/01).

Hydrolyzed: See Hydrolysis.

Hydrophilic/Hydrophobic: A thermodynamically favored interaction between a substance and water (2011/01/01).

Hydrophobic interaction: The hydrophobic interaction is mostly an entropic effect originating from the disruption of the highly dynamic hydrogen bonds between molecules of liquid water by a nonpolar solute.

Hydroxyapatite: Mineral component of bones and teeth (1994/01/01).

Hydroxyl radical: The univalent radical OH. Hydroxyl radical is a potent oxidizing agent (1994/01/01).

Hypoxia: A reduction in brain oxygen supply due to anoxemia (a reduced amount of oxygen being carried in the blood by hemoglobin), or to a restriction of the blood supply to the brain or both (1999/01/01).

ICI: Imperial Chemical Industries.

Immunohistochemical test: A test based on an immunochemistry method. See Immunohistochemistry.

Immunohistochemistry: Histochemical localization of immunoreactive substances using labeled antibodies as reagents (1988/01/01).

***In Silico*:** A Latin word often used for 'computer-based representation' of physical systems and phenomena such as chemical processes (1987/01/01). However, this is an improper use of the Latin dative form of *in silico* as the computer is based on silicium and not silica chips.

Induced pluripotent stem cell: Cell from adult organisms that have been reprogrammed into a pluripotential state similar to that of embryonic stem cells (2010/01/01).

Interferon: Proteins secreted by vertebrate cells in response to a wide variety of inducers. They confer resistance against many different viruses, inhibit proliferation of normal and malignant cells, impede multiplication of intracellular parasites, enhance macrophage and granulocyte phagocytosis, augment natural killer cell activity and show several other immunomodulatory functions (1983/01/01).

Interleukin: Soluble factors which stimulate growth-related activities of leukocytes as well as other cell types. They enhance cell proliferation and differentiation, DNA synthesis, secretion of other biologically active molecules and responses to immune and inflammatory stimuli (1988/01/01).

Connective tissue: Tissue that supports and binds other tissues. It consists of connective tissue cells embedded in a large amount of extracellular matrix (1966/01/01).

Intravasation: Intravasation is the invasion of cancer cells through the basement membrane into a blood or lymphatic vessel.

Ionic interaction: A type of chemical bonding that involves the electrostatic attraction between oppositely charged ions, or between two atoms with sharply different electronegativities, and is the primary interaction occurring in ionic compounds.

Ionic strength = osmolar concentration: The concentration of osmotically active particles in solution expressed in terms of osmoles of solute per liter of solution. Osmolality is expressed in terms of osmoles of solute per kilogram of solvent (1970/01/01).

Ionizing radiations: Electromagnetic radiation or particle radiation (high energy elementary particles) capable of directly or indirectly producing IONS in its passage through matter. The wavelengths of ionizing electromagnetic radiation are equal to or smaller than those of short (far) ultraviolet radiation and include gamma and X-rays (1976/01/01).

iPS cell: Induced pluripotent stem cell.

Keratinocytes: Epidermal cells which synthesize keratin and undergo characteristic changes as they move upward from the basal layers of the epidermis to the cornified (horny) layer of the skin. Successive stages of differentiation of the keratinocytes forming the epidermal layers are basal cell, spinous or prickle cell, and the granular cell (1990/01/01).

Kinases or phosphotransferases: A rather large group of enzymes comprising not only those transferring phosphate but also diphosphate, nucleotidyl residues and others. These have also been sub-divided according to the acceptor group. (From *Enzyme Nomenclature*, 1992) EC 2.7.x.x. (1963/01/01).

KRAS: An oncogene of the RAS family (Kirsten-RAS), a member of the family of Retrovirus-Associated DNA Sequences (RAS). All genes of the family have a similar exon-intron structure and each encodes a p21 protein (1988/01/01).

Kupffer cells: Specialized phagocytic cells of the mononuclear phagocyte system found on the luminal surface of the hepatic sinusoids. They filter bacteria and small foreign proteins out of the blood and dispose of worn out red blood cells (1973/01/01). They are also known as stellate macrophages.

LDL: Low-density lipoproteins.

***Legionella*:** Gram-negative aerobic rods, isolated from surface water or thermally polluted lakes or streams. Members are pathogenic for man. *Legionella pneumophila* is the causative agent for Legionnaires' disease (1981/01/01).

Lentivirus: A genus of the family *Retroviridae* consisting of non-oncogenic retroviruses that produce multi-organ diseases characterized by long incubation periods and persistent infection. Lentiviruses are unique in that they contain open reading frames between the pol and env genes and in the 3′ env region. Five serogroups are recognized, reflecting the mammalian hosts with which they are associated. HIV-1 is the type species (1991/01/01).

Leukemia: A progressive, malignant disease of the blood-forming organs, characterized by distorted proliferation and development of leukocytes and their precursors in the blood and bone marrow. Leukemias were originally termed acute or chronic based on life expectancy but now are classified according to cellular maturity. Acute leukemias consist of predominately immature cells; chronic leukemias are composed of more mature cells (From *The Merck Manual*, 2006) (1966/01/01).

Leukocytes: White blood cells. These include granular leukocytes (basophils; eosinophils; and neutrophils) as well as non-granular leukocytes (lymphocytes and monocytes) (1966/01/01).

Ligand: A molecule that binds to another molecule, used especially to refer to a small molecule that binds specifically to a larger molecule, for example, an antigen binding to an antibody, a hormone or neurotransmitter binding to a receptor, or a substrate or allosteric effector binding to an enzyme. Ligands are also molecules that donate or accept a pair of electrons to form a coordinate covalent bond with the central metal atom of a coordination complex (From Dorland, 27th ed.) (1974/01/01).

Lipid peroxidation: Peroxidase catalyzed oxidation of lipids using hydrogen peroxide as an electron acceptor (1989/01/01).

Liposomes: Artificial, single or multilaminar vesicles (made from lecithins or other lipids) that are used for the delivery of a variety of biological molecules or molecular complexes to cells, for example, drug delivery and gene transfer. They are also used to study membranes and membrane proteins (1973/01/01).

Liver sinusoid cells: A liver sinusoid is a type of capillary known as a sinusoidal capillary, discontinuous capillary or sinusoid, that is similar to a fenestrated capillary, having discontinuous endothelium that serves as a location for mixing of the oxygen-rich blood from the hepatic artery and the nutrient-rich blood from the portal vein.

LNP: Lipid Nanoparticles.

Lymphatic system: A system of organs and tissues that process and transport immune cells and lymph (1966/01/01).

Lymphocyte B/T: White blood cells formed in the body's lymphoid tissue. The nucleus is round or ovoid with coarse, irregularly clumped chromatin while the cytoplasm is typically pale blue with azurophilic (if any) granules. Most lymphocytes can be classified as either T or B (with subpopulations of each), or natural killer cells (1966/01/01).

Macrophage: The relatively long-lived phagocytic cell of mammalian tissues that are derived from blood monocytes. The main types are peritoneal macrophages, alveolar macrophages, histiocytes, Kupffer cells of the liver, and osteoclasts. They may further differentiate within chronic inflammatory lesions to epithelioid cells or may fuse to form foreign body giant cells or langhans giant cells (from the *Dictionary of Cell Biology*, Lackie And Dow, 3rd ed.) (1966/01/01).

Macropinocytosis: Bulk uptake of extracellular components using a mechanism different from pinocytosis.

Magnetic resonance imaging (MRI): Is a medical imaging technique used in radiology to form pictures of the internal anatomy and the physiological processes of the body. MRI scanners use strong magnetic fields, magnetic field gradients and radio waves to generate images of the organs in the body. MRI does not involve X-rays or the use of ionizing radiation, which distinguishes it from CT and PET scans.

Mass spectrometry: An analytical method used in determining the identity of a chemical based on its mass using mass analyzers/mass spectrometers (1974/01/01).

Melanoma: A malignant neoplasm derived from cells that are capable of forming melanin, which may occur in the skin of any part of the body, in the eye or, rarely, in the mucous membranes of the genitalia, anus, oral cavity or other sites. It occurs mostly in adults and may originate de novo or from a pigmented nevus or malignant lentigo. Melanomas frequently metastasize widely, and the regional lymph nodes, liver, lungs and brain are likely to be involved. The incidence of malignant skin melanomas is rising rapidly in all parts of the world (Stedman, 25th ed.; from Rook et al., *Textbook of Dermatology*, 4th ed., p. 2445) (1966/01/01).

Meningioma: A relatively common neoplasm of the central nervous system that arises from arachnoidal cells. The majority are well-differentiated vascular tumors, which grow slowly and have a low potential to be invasive, although malignant subtypes occur. Meningiomas have a predilection to arise from the parasagittal region, cerebral convexity, sphenoidal ridge, olfactory groove and spinal canal (From DeVita et al., *Cancer: Principles and Practice of Oncology*, 5th ed., pp. 2056–7) (1966/01/01).

Mesenchymal stem cell: Also referred to as multipotent stromal cells or mesenchymal stromal cells are multipotent, non-hematopoietic adult stem cells that are present in multiple tissues, including bone marrow, adipose tissue and Wharton's jelly. Mesenchymal stem cells can differentiate into mesodermal lineages, such as adipocytic, osteocytic and chondrocytic cells (2012/01/01) (see fibroblast).

Mesentery: A layer of the peritoneum which attaches the abdominal viscera to the abdominal wall and conveys their blood vessels and nerves (1966/01/01).

Mesothelioma: A tumor derived from mesothelial tissue (peritoneum, pleura, pericardium). It appears as broad sheets of cells, with some regions containing spindle-shaped, sarcoma-like cells and other regions showing adenomatous patterns. Pleural mesotheliomas have been linked to exposure to asbestos (Dorland, 27th ed.) (1966/01/01).

Metabolism: The chemical reactions in living organisms by which energy is provided for vital processes and activities and new material is assimilated (1960/01/01).

Metabolomics: The systematic identification and quantitation of all the metabolic products of a cell, tissue, organ or organism under varying conditions. The metabolome of a cell or organism is a dynamic collection of metabolites which represent its net response to current conditions (2009/01/01).

Metalloprotease: Proteases which use a metal, normally zinc, in the catalytic mechanism. This group of enzymes is inactivated by metal chelators (2004/01/01).

Metastasis: Of a neoplasm. The transfer of a neoplasm from one organ or part of the body to another remote from the primary site (1966/01/01).

MIAME: Minimum Information About a Microarray Experiment.

Micelles: Particles consisting of aggregates of molecules held loosely together by secondary bonds. The surfaces of micelles are usually comprised of amphipathic compounds that are oriented in a way that minimizes the energy of interaction between the micelle and its environment. Liquids that contain large numbers of suspended micelles are referred to as emulsions (1991/01/01).

Microfluidics: The study of fluid channels and chambers of tiny dimensions of tens to hundreds of micrometers and volumes of nanoliters or picoliters. This is of interest in biological microcirculation and used in microchemistry and investigative techniques (2004/01/01).

Microglia cell: The third type of glial cell, along with astrocytes and oligodendrocytes (which together form the macroglia). Microglia vary in appearance depending on developmental stage, functional state and anatomical location; subtype terms include ramified, perivascular, ameboid, resting and activated. Microglia clearly are capable of phagocytosis and play an important role in a wide spectrum of neuropathologies. It has also been suggested that they act in several other roles including in secretion (e.g., of cytokines and neural growth factors), in immunological processing (e.g., antigen presentation) and in central nervous system development and remodeling (1994/01/01).

MINSEQE: Minimum Information About a Next-generation Sequencing Experiment.

MIT: Massachusetts Institute of Technology.

Mitochondrion: Semiautonomous, self-reproducing organelle that occur in the cytoplasm of all cells of most, but not all, eukaryotes. Each mitochondrion is surrounded by a double limiting membrane. The inner membrane is highly invaginated, and its projections are called cristae. Mitochondria are the sites of the reactions of oxidative phosphorylation, which result in the formation of ATP. They contain distinctive ribosomes, and transfer RNAs, amino acyl tRNA synthetases, and elongation and termination factors. Mitochondria depend upon genes within the nucleus of the cells in which they reside for many essential messenger RNAs. Mitochondria are believed to have arisen from aerobic bacteria that established a symbiotic relationship with primitive protoeukaryotes (King & Stansfield, *A Dictionary of Genetics*, 4th ed.) (1960/01/01).

Mitophagy: Proteolytic breakdown of the mitochondria via autophagy (2013/01/01).

Monoclonal antibody: Antibodies produced by a single clone of cells (1982/01/01).

Monocyte: Large, phagocytic mononuclear leukocytes produced in the vertebrate bone marrow and released into the blood; contain a large, oval or somewhat indented nucleus surrounded by voluminous cytoplasm and numerous organelles; mature macrophage derived from blood monocytes (1966/01/01).

Mononuclear phagocyte system: Mononuclear cells with pronounced phagocytic ability that are distributed extensively in lymphoid and other organs. It includes macrophages and their precursors, phagocytes, Kupffer cells, histiocytes, dendritic cells, Langerhans cells and microglia. The term mononuclear phagocyte system has replaced the former reticuloendothelial system, which also included less active phagocytic cells such as fibroblasts and endothelial cells (From *Illustrated Dictionary of Immunology*, 2nd ed.) (1966/01/01).

mRNA: Messenger RNA, sequences that serve as templates for protein synthesis. Bacterial mRNAs are generally primary transcripts in that they do not require post-transcriptional processing. Eukaryotic mRNA is synthesized in the nucleus and must be exported to the cytoplasm for translation. Most eukaryotic mRNAs have a sequence of polyadenylic acid at the 3′ end, referred to as the poly(a) tail. The function of this tail is not known for certain, but it may play a role in the export of mature mRNA from the nucleus as well as in helping stabilize some mRNA molecules by retarding their degradation in the cytoplasm (1965/01/01).

MSC: Mesenchymal Stem Cell.

Mucociliary clearance: A non-specific host defense mechanism that removes mucus and other material from the lungs by ciliary and secretory activity of the tracheobronchial submucosal glands. It is measured in vivo as mucus transfer, ciliary beat frequency and clearance of radioactive tracers (1988/01/01).

Mucosa: A mucous membrane or mucosa is a membrane that lines various cavities in the body of an organism and covers the surface of internal organs. It consists of one or more layers of epithelial cells overlying a layer of loose connective tissue.

Multiple sclerosis: An autoimmune disorder mainly affecting young adults and characterized by destruction of myelin in the central nervous system. Pathologic findings include multiple sharply demarcated areas of demyelination throughout the white matter of the central nervous system. Clinical manifestations include visual loss, extra-ocular movement disorders, paresthesias, loss of sensation, weakness, dysarthria, spasticity, ataxia and bladder dysfunction. The usual pattern is one of recurrent attacks followed by partial recovery, but acute fulminating and chronic progressive forms also occur (Adams et al., *Principles of Neurology*, 6th ed., p. 903), (1966/01/01).

Multipotent: See Stem Cells.

***Mycobacterium tuberculosis*:** A species of gram-positive, aerobic bacteria that produces tuberculosis in humans, other primates, cattle; dogs; and some other animals which have contact with humans. Growth tends to be in serpentine, cordlike masses in which the bacilli show a parallel orientation (1966/01/01).

Myelin: The lipid-rich sheath surrounding axons in both the central nervous systems and the peripheral nervous system. The myelin sheath is an electrical insulator and allows faster and more energetically efficient conduction of impulses. The sheath is formed by the cell membranes of glial cells (Schwann cells in the peripheral and oligodendroglia in the central nervous system). Deterioration of the sheath in demyelinating diseases is a serious clinical problem (1966/01/01).

Myocyte: Mature contractile cells, commonly known as myocytes, that form one of three kinds of muscles. The three types of muscle cells are skeletal (muscle fibers, skeletal), cardiac (myocytes, cardiac) and smooth (myocytes, smooth muscle). They are derived from embryonic (precursor) muscle cells called myoblasts (2003/01/01).

Myristates: Salts and esters of the 14-carbon saturated monocarboxylic acid-myristic acid (1991/01/01).

Nanoelectronics: Refers to the use of nanotechnology in electronic components.

Nanomedicine: The branch of medicine concerned with the application of nanotechnology to the prevention and treatment of disease. It involves the monitoring, repair, construction and control of human biological systems at the molecular level, using engineered nanodevices and nanostructures (From Freitas Jr., *Nanomedicine*, vol. 1, 1999) (2006/01/01). In this book, we have construed nanomedicine as widely as possible, potentially including all the contributions of the nanosciences and nanotechnology to medicine. This being said, it is very difficult for anyone to say where the nanosciences end and nanotechnology begins or vice versa.

Nanoparticle: Nanometer-sized particles that are nanoscale in three dimensions. They include nanocrystalline materials, nanocapsules, metal nanoparticles, dendrimers and quantum dots. The uses of nanoparticles include drug delivery systems and cancer targeting and imaging (2007/01/01).

Nanoscale or nanoscopic scale: Refers to structures with a length scale applicable to nanotechnology, usually cited as 1–100 nm.

Nanosensor: Nanoscale device that measures a physical quantity and converts it to a signal.

Nanotechnology: The development and use of techniques to study physical phenomena and construct structures in the nanoscale size range or smaller (2002/01/03).

Nanotube: Nanometer-sized tubes composed of various substances including carbon (carbon nanotubes, CNT), boron nitride or nickel vanadate (2004/01/01).

NBIC: Nanotechnology, Biotechnology, Information technology and Cognitive science.

***Neisseria meningitidis*:** A species of gram-negative, aerobic bacteria. It is a commensal and pathogen only of humans, and can be carried asymptomatically in the nasopharynx. When found in cerebrospinal fluid it is the causative agent of cerebrospinal meningitis (1966/01/01).

Neurite: In tissue culture, hairlike projections of neurons stimulated by growth factors and other molecules. These projections may go on to form a branched tree of dendrites or a single axon or they may be reabsorbed at a later stage of development. 'Neurite' may refer to any filamentous or pointed outgrowth of an embryonal or tissue-culture neural cell (1992/01/01).

Neuroblastoma: A common neoplasm of early childhood arising from neural crest cells in the sympathetic nervous system, and characterized by diverse clinical behavior, ranging from spontaneous remission to rapid metastatic progression and death. This tumor is the most common intra-abdominal malignancy of childhood, but it may also arise from thorax and neck, or rarely occur in the central nervous system (1966/01/01).

Neuroendocrine tumors: Tumors whose cells possess secretory granules and originate from the neuroectoderm, that is, the cells of the ectoblast or epiblast that program the neuroendocrine system. Common properties across most neuroendocrine tumors include ectopic hormone production (often via Apud cells), the presence of tumor-associated antigens and isozyme composition (1994/01/01).

Neurotrophic factor: Factor that enhances the growth potentialities of sensory and sympathetic nerve cells. An example is the Nerve Growth Factor (NGF) (1972/01/01).

Neutropenia: A decrease in the number of neutrophils found in the blood (1991/01/01).

Neutrophils: Granular leukocytes having a nucleus with three to five lobes connected by slender threads of chromatin, and cytoplasm containing fine inconspicuous granules and stainable by neutral dyes (1965/01/01).

NFκB: Nuclear factor kappa-light-chain-enhancer of B-cells.

nm: Nanometer. One billionth of a meter. One thousandth of a micrometer.

NP: Nanoparticle.

Nubot: Nanorobot.

Nuclear factor kappa-light-chain-enhancer of B-cells: Ubiquitous, inducible, nuclear transcriptional activator that binds to enhancer elements in many different cell types and is activated by pathogenic stimuli. The Nf-Kappa B complex is a heterodimer composed of two DNA-binding subunits, Nf-Kappa B1 and RELA (1991/01/01).

Nucleic acids: High-molecular-weight polymers containing a mixture of purine and pyrimidine nucleotides chained together by ribose or deoxyribose linkages (1966/01/01). Nuclein was the original denomination of nucleic acids given by Friedrich Miescher.

Nucleotide: The monomeric units from which DNA or RNA polymers are constructed. They consist of a purine or pyrimidine base, a pentose sugar and a phosphate group (From King & Stansfield, *A Dictionary of Genetics*, 4th ed.) (1966/01/01).

Octafluoropropane: Octafluoropropane (C_3F_8) is the perfluorocarbon counterpart to the hydrocarbon propane.

Oligonucleotides: Polymers made up of a few (2–20) nucleotides. In molecular genetics, they refer to a short sequence synthesized to match a region where a mutation is known to occur, and then used as a probe (oligonucleotide probes) (1974/01/01).

Oligopeptide: Peptides composed of between two and 12 amino acids (1973/01/01).

Omics: Various disciplines in biology whose names end in the suffix -omics, such as genomics, proteomics, metabolomics, metagenomics, phenomics and transcriptomics. Omics aims at the collective characterization and quantification of pools of biological molecules that translate into the structure, function and dynamics of an organism or organisms.

Oncogenes: Genes whose gain-of-function alterations lead to neoplastic cell transformation. They include, for example, genes for activators or stimulators of cell proliferation such as growth factors, growth factor receptors, protein kinases, signal transducers, nuclear phosphoproteins and transcription factors. A prefix of 'v-' before oncogene symbols indicates oncogenes captured and transmitted by retroviruses; the prefix 'c-' before the gene symbol of an oncogene indicates it is the cellular homolog (proto-oncogenes) of a v-oncogene (1983/01/01).

Organelles: Specific particles of membrane-bound organized living substances present in eukaryotic cells, such as the mitochondria; the Golgi apparatus, endoplasmic reticulum, lysosomes, plastids and vacuoles (1989/01/01).

Organoids: An organization of cells into an organ-like structure. Organoids can be generated in culture. They are also found in certain neoplasms (1965/01/01).

Osteoblasts: Bone-forming cells which secrete an extracellular matrix. Hydroxyapatite crystals are then deposited into the matrix to form bone (1965/01/01).

Osteoclasts: Large multinuclear cell associated with the bone resorption (1963/01/01).

Osteocytes: Mature osteoblasts that have become embedded in the bone matrix. They occupy a small cavity, called lacuna, in the matrix and are connected to adjacent osteocytes via protoplasmic projections called canaliculi (1965/01/01).

Osteofibrosis = primary myelofibrosis: A de novo myeloproliferation arising from an abnormal stem cell. It is characterized by the replacement of bone marrow by fibrous tissue, a process that is mediated by cytokines arising from the abnormal clone (2009/01/01).

Osteolysis: Dissolution of bone that particularly involves the removal or loss of calcium (1987/01/01).

PA: Peptide amphiphile.

Paclitaxel: A cyclodecane isolated from the bark of the Pacific yew tree, *Taxus Brevifolia*. It stabilizes microtubules in their polymerized form leading to cell death (1993/01/01).

Paracetamol = acetaminophen: Analgesic and antipyretic drug.

Paracrine communication: Cellular signaling in which a factor secreted by a cell affects other cells in the local environment. This term is often used to denote the action of intercellular signaling peptides and proteins on surrounding cells (1998/01/01).

Paramagnetic material: A material that displays paramagnetism: a form of magnetism whereby some materials are weakly attracted by an externally applied magnetic field, and form internal, induced magnetic fields in the direction of the applied magnetic field.

Parkinson's disease: A progressive, degenerative neurologic disease characterized by a tremor that is maximal at rest, retropulsion (i.e., a tendency to fall backward), rigidity, stooped posture, slowness of voluntary movements and a masklike facial expression. Pathologic features include loss of melanin containing neurons in the substantia nigra and other pigmented nuclei of the brainstem. Lewy bodies are present in the substantia nigra and locus coeruleus but may also be found in a related condition characterized by dementia in combination with varying degrees of parkinsonism (Adams et al., Principles of Neurology, 6th ed., p. 1059, pp. 1067–75) (1967/01/01).

PEG: Polyethylene glycols, polymers of ethylene oxide and water, and their ethers.

PEGylated: Made with the addition of PEG molecules.

Peptide amphiphile: Peptide-based molecule with both hydrophobic and hydrophilic characters that self-assembles into a supramolecular nanostructure.

PET: Positron Emission Tomography. A medical imaging technique based on the detection of positrons emitted from the body.

pH = hydrogen-ion concentration: The normality of a solution with respect to hydrogen ions, H+. It is related to acidity measurements in most cases by pH = log 1/2[1/(H+)], where (H+) is the hydrogen ion concentration in gram equivalents per liter of solution (McGraw-Hill *Dictionary of Scientific and Technical Terms*, 6th ed.) (1966/01/01).

Phagocytosis: The engulfing and degradation of microorganisms; other cells that are dead, dying or pathogenic; and foreign particles by phagocytic cells (phagocytes) (1966/01/01).

Phagosome: Membrane-bound cytoplasmic vesicles formed by invagination of phagocytized material. They fuse with lysosomes to form phagolysosomes in which the hydrolytic enzymes of the lysosome digest the phagocytized material (1986/01/01).

Pharmacodynamics: The study of the biochemical and physiologic effects of drugs (especially pharmaceutical drugs). The effects can include those manifested within animals (including humans), microorganisms or combinations of organisms (e.g., infection).

Pharmacokinetics: Dynamic and kinetic mechanisms of exogenous chemical drug liberation; absorption; biological transport; tissue distribution; biotransformation; elimination; and drug toxicity as a function of dosage and rate of metabolism. LADMER, ADME and ADMET are abbreviations for liberation, absorption, distribution, metabolism, elimination and toxicology (1988/01/01).

Pharmacopoeia: Authoritative work containing lists of drugs and preparations, their description, formulation, analytic composition, main chemical properties, standards for strength, purity, dosage, chemical tests for determining identity, etc. They have the status of a reference and standard for pharmacists (1997/01/01).

Pharmacovigilance: The detection of long- and short-term side effects of conventional and traditional medicines through research, data mining, monitoring and evaluation of healthcare information obtained from healthcare providers and patients (2012/01/01).

Pharyngeal: See Pharynx.

Pharynx: A funnel-shaped fibromuscular tube that conducts food to the esophagus, and air to the larynx and lungs (1966/01/01).

Phosphocholine: Phosphocholine is an intermediate in the synthesis of phosphatidylcholine in tissues.

Phospholipids: Lipids containing one or more phosphate groups, particularly those derived from either glycerol (phosphoglycerides see glycerophospholipids) or sphingosine (sphingolipids). They are polar lipids that are of great importance for the structure and function of cell membranes and are the most abundant of membrane lipids, although not stored in large amounts in the system (1960/01/01).

Phosphorylation: The introduction of a phosphoryl group into a compound through the formation of an ester bond between the compound and a phosphorus moiety (1991/01/01).

Photolithography = optical lithography: A process using ultraviolet light to transfer a geometric design from an optical mask to a light-sensitive chemical (photoresist) coated on the substrate.

Photonic crystal: An optical nanostructure in which the refractive index changes periodically.

Pinocytosis: The engulfing of liquids by cells by a process of invagination and closure of the cell membrane to form fluid-filled vacuoles (1966/01/01).

Plasmids: Extrachromosomal, usually circular, DNA molecules that are self-replicating and transferable from one organism to another. They are found in a variety of bacterial, archaeal, fungal, algal and plant species. They are used in genetic engineering as cloning vectors (1978/01/01).

PLGA: Polylactic acid–polyglycolic acid copolymer.

Pluripotent: See Stem Cells.

Pneumocyte: Or alveolar epithelial cells. Epithelial cells that line the pulmonary alveoli (2010/01/01).

Polyethylenimine: Strongly cationic polymer that binds to certain proteins; used as a marker in immunology, to precipitate and purify enzymes and lipids (1991/01/01).

Polylactic acid–polyglycolic acid copolymer: A co-polymer that consists of varying ratios of polylactic acid and polyglycolic acid. It is used as a matrix for drug delivery and for bone regeneration (2019/01/01).

Polymer: A substance or material consisting of very large molecules, or macromolecules, composed of many repeating subunits. Due to their broad spectrum of properties, both synthetic and natural polymers play essential and ubiquitous roles in everyday life.

Polysomes: A multiribosomal structure representing a linear array of ribosomes held together by messenger RNA (RNA, messenger); they represent the active complexes in cellular protein synthesis and are able to incorporate amino acids into polypeptides both in vivo and in vitro (From Rieger et al., Glossary of Genetics, Classical and Molecular, 5th ed.) (1973/01/01).

Positron emission tomography: An imaging technique using compounds labeled with short-lived positron-emitting radionuclides (such as carbon-11, nitrogen-13, oxygen-15 and fluorine-18) to measure cell metabolism. It has been useful in study of soft tissues such as cancer, cardiovascular system and brain. single-photon emission-computed tomography is closely related to positron emission tomography, but uses isotopes with longer half-lives and resolution is lower (2005/01/01).

Preclinical developments: In drug development, preclinical development, also termed preclinical studies or nonclinical studies, is a stage of research that begins before clinical trials (testing in humans) and during which important feasibility, iterative testing and drug safety data are collected, typically in laboratory animals.

Precision medicine: Clinical, therapeutic and diagnostic approaches to optimal disease management based on individual variations in a patient's genetic profile (2010/01/01).

Probe (molecular): A group of atoms or molecules attached to other molecules or cellular structures and used in studying the properties of these molecules and structures. Radioactive DNA or RNA sequences are used in molecular genetics to detect the presence of a complementary sequence by nucleic acid hybridization (1989/01/01).

Progesterone: The major progestational steroid that is secreted primarily by the corpus luteum and the placenta. Progesterone acts on the uterus, the mammary glands and the brain. It is required in embryo implantation, pregnancy maintenance and the development of mammary tissue for milk production. Progesterone, converted from pregnenolone, also serves as an intermediate in the biosynthesis of gonadal steroid hormones and adrenal corticosteroids (1966/01/01).

Propranolol: A widely used non-cardioselective beta-adrenergic antagonist. Propranolol has been used for myocardial infarction; arrhythmia; angina pectoris; hypertension; hyperthyroidism; migraine; pheochromocytoma and anxiety but adverse effects instigate replacement by newer drugs (1967/01/01).

Prostate-specific antigen: A glycoprotein that is a kallikrein-like serine proteinase and an esterase—hK3 Kallikrein—produced by epithelial cells of both normal and malignant prostate tissue. It is an important marker for the diagnosis of prostate cancer (1993/01/01).

Protein translation or protein biosynthesis: The biosynthesis of peptides and proteins on ribosomes, directed by messenger RNA, via transfer RNA that is charged with standard proteinogenic amino acids (1973/01/01).

Proteomics: The systematic study of the complete complement of proteins (proteome) of organisms (2003/01/01).

PSA: Prostate-specific antigen.

Pulmonary alveoli: Small polyhedral outpouchings along the walls of the alveolar sacs, alveolar ducts and terminal bronchioles through the walls of which gas exchange between alveolar air and pulmonary capillary blood takes place (1966/01/01).

PVC: Polyvinyl chloride.

QR: Quick response.

QSAR: Quantitative structure–activity relationship.

Quantitative structure–activity relationship: A quantitative prediction of the biological, ecotoxicological or pharmaceutical activity of a molecule. It is based upon structure and activity information gathered from a series of similar compounds (2001/01/01).

Quantum dot: Nanometer-sized fragments of semi-conductor crystalline material which emit photons. The wavelength is based on the quantum confinement size of the dot. They can be embedded in microbeads for high throughput analytical chemistry techniques (2004/01/01).

Radioactive decay = radioactivity: The spontaneous transformation of a nuclide into one or more different nuclides, accompanied by either the emission of particles from the nucleus, nuclear capture or ejection of orbital electrons, or fission (*McGraw-Hill Dictionary of Scientific and Technical Terms*, 6th ed.) (1966/01/01).

Radioisotopes: Isotopes that exhibit radioactivity and undergo radioactive decay (From Grant & Hackh's Chemical Dictionary, 5th ed. & McGraw-Hill Dictionary of Scientific and Technical Terms, 4th ed.) (1966/01/01).

Reactive oxygen species: Molecules or ions formed by the incomplete one-electron reduction of oxygen. These reactive oxygen intermediates include singlet oxygen; superoxides; peroxides; hydroxyl radical and hypochlorous acid. They contribute to the microbicidal activity of phagocytes, regulation of signal transduction and gene expression, and the oxidative damage to nucleic acids; proteins and lipids (1993/01/01).

Replicon: Any DNA sequence capable of independent replication or a molecule that possesses a replication origin and which is therefore potentially capable of being replicated in a suitable cell (Singleton & Sainsbury, *Dictionary of Microbiology and Molecular Biology*, 2nd ed.) (1980/01/01).

Reverse transcription: The biosynthesis of DNA carried out on a template of RNA (2005/01/01).

RGD peptide: Arginylglycylaspartic acid (RGD peptide) is the most common peptide motif responsible for cell adhesion to the extracellular matrix (ECM).

Ribonucleotide: Nucleotides in which the purine or pyrimidine base is combined with ribose, the unitary motive of RNA (1973/01/01).

Ribosome: Multicomponent ribonucleoprotein structures found in the cytoplasm of all cells, and in mitochondria, and plastids. They function in protein biosynthesis via genetic translation (1965/01/01).

Richard phillips feynman: (1918–1988) An American theoretical physicist, known for his work in the path integral formulation of quantum mechanics, and Nobel prize winner in 1965. He presented certain principles of nanotechnology in a talk given in 1959 called 'There's Plenty of Room at the Bottom'.

RNAses: Ribonucleases.

ROS: Reactive oxygen species.

saRNA: Self-amplifying RNA.

SARS-CoV-2 Virus: The coronavirus responsible of severe acute respiratory syndrome (COVID-19 disease).

Scanning tunneling microscopy: A type of scanning probe microscopy in which a very sharp conducting needle is swept just a few angstroms above the surface of a sample. The tiny tunneling current that flows between the sample and the needle tip is measured, and from this are produced three-dimensional topographs. Due to the poor electron conductivity of most biological samples, thin metal coatings are deposited on the sample (1991/01/01).

Single nucleotide individual polymorphism: A single nucleotide variation in a genetic sequence that occurs at appreciable frequency in the population (2000/01/01).

Single-stranded DNA: A single chain of deoxyribonucleotides that occurs in some bacteria and viruses but may also by synthetized (1974/01/01).

siRNA: Small interfering RNA.

Small interfering RNA: Small double-stranded, non-protein coding RNA (21–31 nucleotides) involved in gene silencing functions, especially RNA interference (RNAi). Endogenously, siRNA are generated from dsRNA (RNA, double-stranded) by the same ribonuclease, dicer, that generates miRNAs (microRNA). The perfect match of the siRNAs' antisense strand to their target RNA mediates RNAi by siRNA-guided RNA cleavage. siRNAs fall into different classes/and have different specific gene silencing functions (2003/01/01).

SNIP: Single nucleotide individual polymorphism.

Solute: A substance dissolved in another substance, known as a solvent.

Somatostatin: A 14-amino acid peptide named for its ability to inhibit pituitary growth hormone release, also called somatotropin release-inhibiting factor (SRIF). It is expressed in the central and peripheral nervous systems, the gut and other organs. SRIF can also inhibit the release of thyroid-stimulating hormone; prolactin; insulin and glucagon besides acting as a neurotransmitter and neuromodulator. In a number of species including humans, there is an additional form of somatostatin, SRIF-28 with a 14-amino acid extension at the N-terminal (1975/09/01).

SPION: Superparamagnetic iron oxide nanoparticles.

Stem cell: Undifferentiated or partially differentiated cells that can differentiate into various types of cells and proliferate indefinitely; they can be totipotent or multipotent and differentiate in all kinds of tissues or only in a few ones. Pluripotent refers only to artificially iPS cells.

Stereolithography: A 3D printing technology where a computer-controlled moving laser beam is used to build up the required structure, layer by layer, from liquid polymers that harden on contact with laser light (photopolymerization) (2018/01/01).

Stressors (chemical): A stressor is a chemical agent causing stress to an organism.

Stroma: The connective, functionally supportive framework of a biological cell, tissue or organ.

Stromal cell: Connective tissue cells of an organ found in the loose connective tissue. These are most often associated with the uterine mucosa and the ovary as well as the hematopoietic system and elsewhere (1993/01/01).

Subacute exposure: An exposure lasting more than 24 hours.

Superoxide: Highly reactive compounds produced when oxygen is reduced by a single electron (1984/01/01).

Symptoms: Signs and symptoms are the observed or detectable signs, and experienced symptoms of an illness, injury or medical condition.

Syncytium or giant cells: Multinucleated masses produced by the fusion of many cells; often associated with viral infections. In AIDS, they are induced when the envelope glycoprotein of the HIV binds to the CD4 antigen of uninfected neighboring T4 cells. The resulting syncytium leads to cell death and thus may account for the cytopathic effect of the virus (1990/01/01).

Syndrome: A characteristic symptom complex (1991/01/01).

T-lymphocytes: Lymphocytes responsible for cell-mediated immunity. Two types have been identified—cytotoxic and helper T-lymphocytes. They are formed when lymphocytes circulate through the thymus gland and differentiate to thymocytes. When exposed to an antigen, they divide rapidly and produce large numbers of new T cells sensitized to that antigen (1973/01/01).

TGF-beta 1: Transforming growth factor beta 1.

Theranostics: Clinical, therapeutic and diagnostic approaches to optimal disease management based on individual variations in a patient's genetic profile (2010/01/01).

Therapeutic index: An indicator of the benefits and risks of treatment (2018/01/01).

Theseus' ship: A philosophical parable. If one imagines a ship in a harbor that has every single plank replaced one by one. At what point (if any) is it no longer the same ship but a new one?

Thrombin: An enzyme formed from prothrombin that converts fibrinogen to fibrin (1966/01/01).

TLR: Toll-like receptor.

Toll-like receptors: A family of pattern recognition receptors characterized by an extracellular leucine-rich domain and a cytoplasmic domain that share homology with the interleukin 1 receptor and the *Drosophila* toll protein. Following pathogen recognition, toll-like receptors recruit and activate a variety of signal transducing adaptor proteins (2006/01/01).

Totipotent stem cells: Single cells that have the potential to form an entire organism. They have the capacity to specialize into extraembryonic membranes and tissues, the embryo, and all postembryonic tissues and organs (2003/01/01).

Toxicant: A toxicant is any toxic substance. Toxicants can be poisonous and they may be man-made or naturally occurring.

Toxicokinetics: The quantitation of the body's metabolism of toxic xenobiotic compounds, as measured by the plasma concentration of the toxicant at various time points (2015/01/01).

Transcription (DNA): Transcription is the process of copying a segment of DNA into RNA.

Transcriptome: The pattern of gene expression at the level of genetic transcription in a specific organism or under specific circumstances in specific cells (2012/01/01).

Transcriptomics: The study of the transcriptome.

Transfection: The uptake of naked or purified DNA by cells, usually meaning the process as it occurs in eukaryotic cells. It is analogous to bacterial transformation and both are routinely employed in gene transfer techniques (1978/01/01).

Transferrin: An iron-binding beta 1 globulin that is synthesized in the liver and secreted into the blood. It plays a central role in the transport of iron throughout the circulation. A variety of transferrin isoforms exist in humans, including some that are considered markers for specific disease states (1966/01/01).

Transforming growth factor beta 1: Transforming growth factor beta 1, a hormonally active polypeptide that can induce the transformed phenotype when added to normal, non-transformed cells (2007/01/01).

Transition metal: An element in the d-block of the periodic table, which includes groups 3 to 12 on the periodic table.

Trastuzumab: A humanized monoclonal antibody against the ERBB-2 receptor (HER2). As an antineoplastic agent, it is used to treat breast cancer where HER2 is overexpressed (2016/01/01).

Triglyceride: An ester formed from glycerol and three fatty acid groups (1968/01/01).

Ultracentrifugation: Centrifugation with a centrifuge that develops centrifugal fields of more than 100,000 times the force of gravity (McGraw-Hill *Dictionary of Scientific and Technical Terms*, 4th ed.), (1965/01/01).

Vaccine: Suspensions of killed or attenuated microorganisms (bacteria, viruses, fungi, protozoa), antigenic proteins, synthetic constructs or other bio-molecular derivatives, administered for the prevention, amelioration or treatment of infectious and other diseases (1966/01/01).

Vacuoles: Any spaces or cavities within a cell. They may function in digestion, storage, secretion or excretion (1989/01/01).

Van der Waal's interactions: In molecular physics, the van der Waals force, named after Dutch physicist Johannes Diderik van der Waals, is a distance-dependent interaction between atoms or molecules.

Vascular endothelial growth factor: The original member of the family of endothelial cell growth factors referred to as vascular endothelial growth factors. Vascular Endothelial Growth Factor A was originally isolated from tumor cells and referred to as 'tumor angiogenesis factor' and 'vascular permeability factor'. Although expressed at high levels in certain tumor-derived cells, it is produced by a wide variety of cell types (2004/01/01).

Vectorization (drug): Targeting specific cells with a drug by means of a nanoparticle.

VEGF: Vascular endothelial growth factor.

Xenobiotic: Means literally 'foreign to life' and refers to all the non-living intruders in the body. Chemical substances that are foreign to the biological system. They include naturally occurring compounds, drugs, environmental agents, carcinogens and insecticides (1989/01/01).

Xenograft: Tissues, cells or organs transplanted between animals of different species (2014/01/01).

Zanamivir: A guanido-neuraminic acid that is used to inhibit neuraminidase (2007/01/01).

Zwitterionic: Molecules possessing an equal number of positively and negatively charged functional groups.

Index

L

M

N

O

Printed in the United States
by Baker & Taylor Publisher Services